The Emancipatory City?

The Emancipatory City?

Paradoxes and Possibilities

edited by
LORETTA LEES

SAGE Publications
London ● Thousand Oaks ● New Delhi

Contents

About the contributors

Ash Amin is Professor and Head of Geography at Durham University. His latest books include: *Cities: Reimagining the Urban* (with Nigel Thrift, 2002, Polity), *Placing the Social Economy* (with Angus Cameron and Ray Hudson, 2002, Routledge), *Architectures of Knowledge* (with Patrick Cohendet, 2004, Oxford University Press) and *The Blackwell Cultural Economy Reader* (edited with Nigel Thrift, 2004, Blackwell).

Les Back is a Reader in the Centre for Urban and Community Research at Goldsmiths College, University of London. His research interests focus on the culture of racism with particular reference to social identity and popular culture. He has published *New Ethnicities and Urban Culture: Racisms and Multiculture in Young Lives* (1995, UCL Press).

Gary Bridge is Senior Lecturer in the Centre for Urban Studies, School for Policy Studies, University of Bristol. He writes on gentrification, class and neighbourhood change, rationality, social action and the city. Recent publications include *A Companion to the City* (2000, Blackwell) and *The Blackwell City Reader* (2002) both co-edited with Sophie Watson. Gary is currently completing a book on philosophical pragmatism and the city – *Reason in the City of Difference* (forthcoming, Routledge).

Gavin Brown has worked as a university administrator for the past decade and is currently researching for a PhD in the Geography Department at King's College London. His current research builds upon earlier academic and applied work on the social and cultural geographies of East London's queer population(s). Central to his research is a desire to engender new activist strategies and his academic work is complemented by an involvement in community politics in East London.

James DeFilippis is Assistant Professor in the Department of Black and Hispanic Studies at City University New York, USA. He is the author of *Unmaking Goliath: Community Control in the Face of Capital Mobility* (2003, Routledge). He has also published on housing, community development, local economic development, urban politics, gentrification and public space.

Nicholas Fyfe teaches geography at the University of Dundee. His particular research interests are in crime and the criminal justice system and, more recently, in the role of voluntary organisations and volunteering in cities in the context of contemporary welfare reforms. He is the editor of *Images of the Street: Planning, Identity and Control in Public Space* (1998, Routledge), author of *Protecting Intimidated Witnesses* (2001, Ashgate), and co-editor with Judith Kenny of *The Urban Geography Reader* (forthcoming, Routledge).

Matthew Gandy teaches geography at University College London and has published widely on cultural, urban and environmental themes. He is author of *Concrete and Clay: Reworking Nature in New York City* (2002, The MIT Press) and co-editor of *The Return of the White Plague: Global Poverty and the 'New' Tuberculosis* (2003, Verso). His current work is focused on cultural histories of urban infrastructure with research in Germany, India, Nigeria and the USA.

David Harvey is Distinguished Professor of Anthropology at the Graduate Center of the City University of New York, USA. He formerly held professorships of geography at Oxford and The Johns Hopkins University. His most recent books include *Justice, Nature and the Geography of Difference* (1996, Blackwell), *Spaces of Hope* (2000, Edinburgh University Press), *Paris, Capital of Modernity* (2003, Routledge) and *The New Imperialism* (2003, Oxford University Press).

Michael Keith is Director of the Centre for Urban and Community Research at Goldsmiths College, University of London. He has written around issues of multiculture and urbanism and is the author of the forthcoming *After the Cosmopolitan: The Future of the Multicultural City* (forthcoming, Routledge).

Peter North is Lecturer in Geography at the University of Liverpool. His research interests are in community action in cities, urban social movements, and community-based responses to neo-liberalism in the global north and south. His current research interests focus on the ongoing resistance to neoliberalism in Argentina as well as community-based struggles in British cities.

Steve Pile is Reader in Cultural Geography in the Faculty of Social Sciences at the Open University. He is author of *The Body and the City* (1996, Routledge) and co-editor, with Nigel Thrift, of *City A-Z* (2000, Routledge). He has also co-edited (with Kay Anderson, Mona Domosh and Nigel Thrift) *Handbook of Cultural Geography* (2003, Sage) and (with Stephan Harrison and Nigel Thrift) *Patterned Ground: Entanglements of Nature and Culture* (2004, Reaktion). He is currently preparing a book on the more phantasmagoric aspects of city life, *Real Cities* (2005, Sage).

David Pinder is Lecturer in Human Geography at Queen Mary, University of London. His research focuses on cities and cultures, and on the utopian visions and urban interventions of modernist and avant-garde movements. He is the co-editor of *Cultural Geography in Practice* (2003, Arnold) and author of *Visions of the City: Utopianism, Power and Politics in Twentieth-Century Urbanism* (2005, Edinburgh University Press).

Geraldine Pratt is Professor of Geography at the University of British Columbia, Vancouver, Canada. Recent work on women's labour and feminist theory appears in *Working Feminism* (2004, Edinburgh University Press). She teaches a course with Rose Marie San Juan on the city and cinema, and they have recently published on this theme in *Screen* 43: 3.

Jennifer Robinson is in the Faculty of Social Sciences at The Open University. She is currently working on a post-colonial critique of urban studies. She is concerned to bring together a field divided by conceptualisations of modernity and the practices of development. Other recent work has explored the politics of post-apartheid urbanism, as well as a critique of global and world cities approaches. *The Power of Apartheid* (1996, Butterworth) drew together earlier work on space and power in South African cities.

Susan Ruddick is Associate Professor in the Department of Geography and Program in Planning, University of Toronto, Canada. She has interests in social theory and the social construction of the child, social reproduction, social policy and social identity. She is the author of *Young and Homeless in Hollywood: Mapping Social Identities* (1996, Routledge).

Rose Marie San Juan teaches in the Department of Art History and Visual Art, University of British Columbia, Vancouver, Canada. She is the author of *Rome: A City Out of Print* (2001, University of Minnesota Press) as well as articles on early modern visual culture and urban life; she has co-authored with Geraldine Pratt articles on the representation of virtual technologies in popular film and is currently completing a book-length study on the early modern deployment of images by the Jesuits in their educational institutions and missionary travels.

Quentin Stevens is Lecturer in Planning and Urban Design at The Bartlett School of Planning, University College London. His research interests combine critical social theory, environment-behaviour relations, and urban morphology. His current research projects explore the complexities of public activity in waterfront leisure precincts, the experiential possibilities of building thresholds, and colonial settlements in Antarctica.

Nigel Thrift is Head of the Division of Life and Environmental Sciences at Oxford University. He has published a number of books on cities and related themes recently, including *City A-Z* (co-edited with Steve Pile, 2000, Routledge), *Timespace* (co-edited with Jon May, 2001, Routledge) and *Cities: Reimagining the Urban* (with Ash Amin, 2002, Polity). He also has research interests in the history of time, management knowledges, international finance and nonrepresentational theory.

List of figures

Preface

This book has been some time in the making. The tests and tribulations of the various authors' lives and careers – illnesses, injuries, promotions, moving home, childcare responsibilities, the increasing pressures and time constraints of the academic world – have pushed the publication of this volume much further back than I would have liked, but the result I think has been the better for it.

The 2003 Oxford Amnesty Lecture Series, Divided Cities, kindly gave permission for David Harvey's lecture to be published in this volume. An expanded version is forthcoming in Richard Scholar (ed.) *Divided Cities: Oxford Amnesty Lectures* (OUP). The other chapters in this book have not been previously published.

I would like to thank Ilse Helbrecht, who was a post-doctoral fellow with me in the Department of Geography at the University of British Columbia, Canada, for it was in discussions with her and others at UBC that I first began to think about the themes in this book. Also Steve Pile who urged me to get this book off the ground and Robert Rojek at Sage for sticking with the project as it became massively overdue.

Finally, thanks to David Demeritt, who has borne the brunt of my frustrations with this book and has supported me throughout, and to our daughter Meg who was born sometime after the proposal for this book was accepted by Sage.

Introduction

1 'The Emancipatory City': Urban (Re)Visions

Loretta Lees

The city beckons to the dreamer in us, for in its vastness and diversity lies a world of fantasy, hope, occasional fulfilment and sadness, longing, loneliness, and the lingering possibility for community with our fellow travellers in the mystery of life ... (Chorney, 1990: 2)

This book interrogates the prospects and possibilities of the city as a space or site for emancipation. For the thousands of new arrivals pouring every day into cities like London, Los Angeles, Sydney, Johannesburg and Mumbai, the city offers the same promise of progress and opportunity it always has to poor rural migrants. If intellectuals have sometimes reacted more ambivalently to the perils of that promise, the city, nevertheless, still lies at the heart of many utopian conceptions of democracy, tolerance, and self-realization. 'A city', Richard Sennett recently declared, 'isn't just a place to live, to shop, to go out and have kids play. It's a place that implicates how one derives one's ethics, how one develops a sense of justice, how one learns to talk with and learn from people who are unlike oneself, which is how a human being becomes human' (1989: 83).

Such utopian hopes for the city date back to antiquity, but they are now more urgent than ever before. More than half of the planet's population now live in cities and the proportion is growing rapidly (UN Centre for Human Settlements, 1996). Whether measured by the number and size of the very largest mega-cities of more than ten million souls, the extent of their cultural and economic reach, the density of their interconnection, or the weight of their ecological footprint upon the planet, the scale of contemporary urbanization is unprecedented. This century, even more than the last, will be an urban one in which the city is the measure of the civility and sustainability of society.

It is in this context that avowedly utopian dreams of an emancipatory city are so vital, if the urban future is to be cast with hope untarnished by fear. As contributors to the final part of this volume insist, the resources required are imaginary as well as material, and in this, those on the Left, in particular, have been somewhat wanting. Decades of suburban growth and inner city decline across much of North America and Europe have provided little for those who love the city to cheer about. While this has given urban studies plenty to do in terms of critique, the relentless focus on urban problems has tended to reinforce longstanding narratives of perverse and pathological urbanism. As John Gold noted some time ago, 'The advocacy of alternative urban visions has all but ceased and, indeed, there is little active debate about the future city other than the projection of current doubts and anxieties into the near future' (1985: 92). The prevailing mood in urban studies has become one of doom and gloom, even fatalism.

Urban studies needs a stronger normative and utopian dimension to complement its tradition of diagnostic critique. Urbanists no longer 'plan the ideal city, but come to terms with the good enough city' (Robins, 1991: 11). The decline of utopianism has left what Raymond Williams (1989) called the 'resources of hope' dangerously underdeveloped. As Merrifield notes 'It is progressive urbanists who need to do the toughest thinking of all. Those of us on the Left who yearn for social justice, but who also love cities, find ourselves torn between the tyranny we see around us every day and the thrill that same tyrannical city can sometimes offer' (2000: 485).

Urban utopianism is not entirely moribund, however. This volume engages with a recent resurgence of avowedly utopian thinking about the city (e.g. Baeten, 2002; Harvey, 2000). In recent years, urbanists, architects and planners have begun, once again, to embrace the urban, or at least selective elements of it, as the solution to problems rather than their source (e.g. DETR, 1999). Advocates of so-called 'new urbanism', for example, celebrate more dense and clustered forms of (sub)urban development as a way to counter soulless suburban sprawl and foster a greater sense of place and of community (see McCann, 1995). While critics complain that this neo-traditional appropriation of urban form without urban diversity or ethos will inevitably descend into gated communities of exclusion (e.g. Dowling, 1998; Ellin, 1996), it should not be dismissed out of hand as simply 'anti-urban'. Such pessimism ignores both its politically progressive critique of unrestricted suburban development and the potential for enlarging its conception of the urban.

Much the same could be said about both the ambivalence and the unrealised potential of urban regeneration. In Britain, for example, the Government has committed itself to urban renaissance as part of a wider international discourse of the so-called 'liveable city'. Critics complain that its Urban Task Force report amounts to a charter for exclusionary gentrification (e.g. Butler and Robson, 2001; Lees, 2000). But there is also much for urbanists to applaud in the Government's hope that more and smarter urbanization can help resolve problems of social exclusion and cohesion in a more multicultural and globally inter-connected economy and society as well as environmental problems of habitat destroying suburban sprawl, congestion, air pollution, and greenhouse gas emissions (Lees, 2003).[1]

This kind of official planning discourse about the liveable city feeds off of and informs more academic claims about the emancipatory potential of the sites and sensuous experiences of the city, city living, and the cityscape. In the context of contemporary debates about the cultural politics of identity, many gays and lesbians, feminists, and anti-racists now embrace the city hopefully as a space of tolerance and diversity in which to realize alternative ways of being (Amin et al., 2000; Fincher and Jacobs, 1998; Keith and Pile, 1993; Sandercock, 1998; Wilson, 1991). Iris Marion Young (1990), to name perhaps the most prominent of these theorists of identity politics, has grounded her vision for cultural pluralism in urban spaces of 'unassimilated otherness' and an 'ideal of city life as a vision of social relations affirming group difference'. Likewise radical critics are returning to the utopian traditions of Marxism to reconsider the urban and its contradictory potential as both the material manifestation of expanding capitalist domination and yet also as a creative process whose experience is the seedbed for forging something new (e.g. Harvey, 2000; Merrifield, 2002).

Whereas other contributors to this volume seek to explore and extend emerging ideas about an emancipatory city, my objective in this introductory chapter is to explain where they have come from. The city has long been celebrated as an emancipatory space/place in the social imaginary of the West. The old German saying, 'city air makes men free' (*Stadtluft macht frei*), refers to the fact that during the Middle Ages living in the city (for those to whom entry was granted) usually meant liberation from feudal master-servant relationships. Other ideas about the city as space for freedom, cradle for civilization, and seed-bed for democracy also go back to antiquity (Mumford, 1995: 21). That classical legacy is inscribed into the very etymology of the word city, which comes from the Latin *civitas*, which meant the body of citizens rather than the place or settlement type where they were gathered. There are also important Biblical traces in ideas of the emancipatory city. Figures as diverse as the Puritan founders of Boston and the Massachusetts Colony, the radical poet William Blake, and Brigham Young and the Mormons, found inspiration in the idea of the city as a New Jerusalem.

Despite the great antiquity of such ideas, it is primarily with the transition to capitalism and the rapid urbanization that it drove that the various contemporary strands of thinking about an emancipatory city emerge. As Williams (1983: 56) notes, 'the city as a really distinctive order of settlement, implying a whole different way of life, is not fully established, with its modern implications, until the early nineteenth century', when *the* city became an abstraction distinct from particular cities or forms of settlement.

For the most part discussion of the emancipatory city has been concerned less with the quality of particular places and urban environments, than with the city itself and the emancipatory potential of all cities. In this sense, then, it is steadfastly utopian both in the conventional sense of being concerned with an ideal city rather than an actually existing (or practically achievable) one, and in the literal sense of u-topian, of being focused on no-place in particular. This is different from the heterotopic focus on the potential for real places to be temporarily transformed into *other* places, 'counter-sites ... [in which] all the other real sites that can be found within the culture, are simultaneously represented, contested' (Foucault, 1986: 24). The contrast between the hetero- and u-topic imagination could easily be over-drawn, especially as the terms have sometimes been used interchangeably in recent work. Nevertheless the distinction is useful in highlighting the double-edged quality of emancipatory city talk. Abstraction from the particular and the actual produces the power to imagine better and more just worlds, but the utopian focus on no place raises important questions about how to realize those ideals in some place. David Harvey (2000) insists that the challenge of realizing an emancipatory city is a dialectical one of reconciling utopias of spatial form with those of social process. That may be so, but a prerequisite for any such effort is achieving some greater conceptual clarity about the often divergent ideals at play in debates about the emancipatory city.

To that end I want to highlight three distinct but related strands of thought. Utopian thinking involves not just an articulation of the ideal and the good, but also a diagnosis of present problems and usually some at least implicit sense of the means of overcoming them and of achieving those utopian ideals. While the three utopian strands I consider here share in common an understanding of the modern city as disruptive, they also open up some quite different problems and possibilities for change.

The emancipatory city and its others

It is difficult to overstate the intellectual legacy of early twentieth century social theorists such as Emile Durkheim, Ferdinand Tönnies, Max Weber, and Georg Simmel and their contrast between an urban society full of strangers and strange experiences and village communities in which everyone knows each other. Urbanization, explained the founder of the Chicago School of Sociology, Robert Park (1952: 24), was liberating but also unsettling: 'the peasant who comes to the city to work and to live, is ... emancipated from the control of ancestral custom but, at the same time, he is no longer backed by the collective wisdom of the peasant community.' Louis Wirth (1938: 192) argued that urban growth created a new 'way of life' founded on Enlightenment values: 'The juxtaposition [in the city] of divergent personali-ties and modes of life tends to produce a relativistic perspective and a sense of toleration of difference which may be regarded as prerequisites for ration-ality and which lead toward the secularization of life'. Urbanization has now become so pervasive as to be almost unremarkable, but echoes of these same constitutive oppositions between country and city, described so ably by Williams (1975), are found today in discussions of the emancipatory city. Elizabeth Wilson (2001: 67), for example, maintains that while 'the distinction between town and country, the provinces and the metropolis is less stark than formerly ... rural life is arguably still even more restrictive and lacking in opportunity for the poor (of both sexes) than life in cities'.

Though commonly represented in romantic and pastoral motifs as a spiritual alienation from nature and community, urbanization was also experienced as a material liberation from the realm of biological necessity. The city orchestrated the conquest of nature. As Matthew Gandy notes in this volume, the urbanization of water led to the conquest of disease, squalor, and human misery. It is testimony to the success of this victory over nature that, despite the continuing scandal of an estimated two million annual deaths in the developing world from preventable water-borne diseases (UN Water World Development Report, 2003: 16), academic debate about the emancipatory city has largely disregarded environmental issues.

Future discussion of the emancipatory city must take the matter of nature much more seriously as a subject of both practical policy and theoretical significance. Contemporary planning discourses of urban densification, containment, and growth control are mobilizing environmental considerations to promote more compact forms of urban development (Lees and Demeritt, 1998). While there is much for urbanists to applaud in this talk of sustainable urbanism, it often has an expert-led, technocratic flavour that serves to disempower citizens and discourage their participation in planning their own future. Somewhat different issues about citizenship are raised by the so-called 'animal turn' in human geography (Philo and Wilbert, 2000; Wolch and Emel, 1998). In highlighting the presence of animals in the city, this work raises important questions about how far the cosmopolitan dream of tolerance and rights to the city is to extend. For Whatmore (2003: 156) the move to bring animals into the cosmopolitan city does not just expand the 'liberal figure of the individual rights-bearing person wholesale to a range of non-human creatures' so much as explode this humanist category altogether and in the process raise far reaching questions about what a more relational ethics of being in difference would look like. This post-humanist perspective is the starting point for Susan Ruddick's exploration in Chapter 2 of the monstrous as a potential figure of openness to difference and emancipation in the city.

If the imagined geographies of the city have long been understood through contrast with the country and traditional rural communities, radical socialists, perhaps more than any other group, have tended to take an optimistic view of the 'emancipatory trajectory of urbanization' (Short, 1991: 41–3). With his modernist faith in progress, Karl Marx believed that capitalism could be transcended by the very forces that it unleashed. The city played a two-fold role in Marx's social theory. On the one hand, it was a fragmented material space created by capitalism and serving to discipline the working class for the purpose of making profit. Hence, urban life was associated with alienation, isolation, and fragmentation. On the other hand, however, the city was also the seat and symbol of historical progress. As *The Manifesto of the Communist Party* famously declares:

> The bourgeoisie has subjected the country to the rule of the towns. It has created enormous cities, has greatly increased the urban population as compared with the rural, and has thus rescued a considerable part of the population from the idiocy of rural life. (Marx and Engels, 1968: 39)

This almost instinctive recoil from 'the idiocy of rural life' was 'habitual among the metropolitan socialists of Europe' (Williams, 1975: 50), who at least until Mao's peasant revolution, have traditionally tended to look to the city for the progressive forms of association that they hoped would herald a better future. Although the industrial city disintegrated family ties and traditional allegiances, fostering individual indifference and isolation, Marx also believed that the city would give rise to new emancipatory forms of association by bringing together different people to discover their common class position. As David Pinder discusses in Chapter 7 this same ambivalent optimism was characteristic of the Situationists and other modernist and avantgarde movements before them.

Second wave feminism of the 1960s and 1970s gave the traditional Marxist opposition of the freedoms of the progressive city to the constraints of a conservative countryside a slightly different twist. Betty Friedan's (1963) *The Feminine Mystique* offered a searing critique of the private, domestic space of the suburbs as an unfulfilling prison for women. Continuous with a wider dissatisfaction with the mindless conformity of the *Organization Man* (Whyte, 1956), Friedan's critique resonated with a generation of women seeking independence from the traditional patriarchal roles of mother and housewife through work and equal opportunity to participate in traditionally male sites and spheres of activity. Picking up on their own personal experience of the city, feminist critics argued that urban living was crucial to the rise of feminist politics and the transformation of restrictive Victorian notions of feminine identity (e.g. Stansell, 1987; Walkowitz, 1992). The city promised women economic independence, and related to this cultural, commercial, and sexual freedom from the patriarchal household in which to forge the 'New Woman' of the twentieth century (Wilson, 1991).

These new urban identities and freedoms were explored and celebrated in modernist novels of female self-discovery. As Heron states:

> It is with modernism that women's fiction truly enters and lays claim to the city, thereby claiming new possibilities for women's autonomy ... [the city is] ... the site of women's most transgressive and subversive fictions throughout the century, as a place where family constraints can be cast off and new freedoms

explored, as a place where the knowledge acquired through urban experience not only brings changed perceptions of identity, but inescapably situates the individual within social order. (1993: 2)

Miriam Henderson, in Dorothy Richardson's *Pilgrimage,* was a 'New Woman', rebelling against patriarchy and finding a transcendence of identity in the city (Heron, 1993: 5). In both *Pilgrimage* and Virginia Woolf's *Mrs Dalloway* the city transforms the female subject into a social subject through a sense of urban empathy due to the proximity of her fellow human beings.

More recent feminist research has challenged this simple opposition of an emancipatory city versus oppressive suburbs (e.g. Spigel, 1992). As Jennifer Robinson's discussion in Chapter 10 suggests that opposition presumes a particular Anglo-American urban geography that is not easily transposed to other contexts, like the townships of South Africa. Others, meanwhile, have noted that Friedan's promise of personal fulfilment through the choice of paid employment in the city presumed a largely middle class privilege not available to women forced to sell their labour in order to survive (e.g. Hanson and Pratt, 1995). Such nuances notwithstanding, Wilson (1991: 46) still insists that the suburban ideal 'acted ideologically to debase and delegitimate the pleasures and possibilities of urban life'. Her vision of the relatively restricted horizons of suburban as opposed to inner city living informs both the practice of gentrification (Lees, 1996) and popular culture, as Geraldine Pratt and Rose Marie San Juan's discussion of the film *The Truman Show* points out in Chapter 12.

Emancipatory alienation and the shock of urban experience

Another source drawn on by contemporary thinking about the emancipatory city is the work of nineteenth and early twentieth century artists and intellectuals struggling to appreciate the unsettling qualities of sensory and intersubjective experience possible in the metropolitan and industrial cities emerging at the time. William Wordsworth, for instance, was one of the first to recognize that the concentration of people into the 'great city' of London was producing a new, disorienting experience of estrangement, indeed, even a new kind of metropolitan subject:

How often, in the overflowing streets,
Have I gone forward with the crowd and said
Unto myself, 'The face of every one
That passes me is a mystery!
Thus I have looked, nor ceased to look, oppressed
Until the shapes before my eyes became
A second-sight procession, such as glides

Over still mountains, or appears in dreams.
And all the ballast of familiar life,
The present, and the past; hope, fear; all stays,
All laws of acting, thinking, speaking man
Went from me, neither knowing me, nor known.

(Wordsworth, 1989: book VII)

For Wordsworth, the dizzying and dream-like experience of walking the city was alienating – from the past, from the familiar, from people now constituted as strangers – but also potentially liberating from this 'ballast'. As Williams (1975: 187) notes, this historically emancipating insight into the potential for making 'new kinds of possible order, new kinds of human unity in the transforming experience of the city, appeared, significantly, in the same shock of recognition of a new dimension' of urban anomie, disorientation, and detraditionalization, which elicited 'the more familiar subjective recoil. The objectively uniting and liberating forces were seen in the same activity as the forces of threat, confusion and loss of identity. And this was how, through the next century and a half, the increasingly dominant fact of the city was to be both paradoxically and alternatively interpreted'.

Wordsworth's unnerving urban experience of excitement and estrangement prefigures many of the concerns of Georg Simmel (1995), whose account of the double-sided cognitive and intersubjective response to sensory overload in the city offers one major wellspring for contemporary hopes of an emancipatory city. According to Simmel people coped with the unceasing intensity of urban interaction by developing a blasé attitude of calculated reserve and detachment from others. While communitarians are inclined to bemoan this atomistic 'form of metropolitan life' (p. 38) as a retreat from meaningful human sociability, Simmel insisted that its apparent 'dissociation is in reality only one of its elemental forms of socialization' (p. 38). For Simmel what 'makes the metropolis the locus of this condition [of personal inner and outer freedom]' (p. 40) is the extent to which its residents are estranged from one another and respond individually to each other with reserved indifference. As Wordsworth anticipated, the effect of this indifference is both alienating, producing a subject who outwardly 'appear[s] to small-town folk … as cold and uncongenial' and inwardly feels 'strangeness and repulsion' from deeper or more intimate contact with most others (p. 37). But paradoxically it is also liberating. Combined with the division of labour, urbanization intensified this blasé attitude and freed the individual from 'the trivialities and prejudices which bind the small-town person' (p. 40) and made the city a space of tolerance for individual difference. Of course the communal restrictions of tradition and mutual expectation were also potential sources of support, and so for Simmel it was only 'the obverse of this [urban] freedom that … one never feels as lonely and as deserted as in this metropolitan crush of persons. For here as elsewhere, it is by no means necessary that the freedom of man reflects itself ... only as a pleasant experience' (p. 40).

Simmel's work informs two distinct but related strands of utopian thinking about the emancipatory city. First, Simmel's (1995: 40) account of the metropolis as the 'seat of cosmopolitanism' is a major influence on the hopes of contemporary feminists, anti-racists, and queer theorists for the city as a potentially emancipatory space of diversity and tolerance. As Pratt and Hanson (1994: 6) argue a 'striking characteristic of many feminist attempts to rethink subjectivity, difference, and political community is their reliance upon place and geography, both as metaphor and material context, as vehicles for doing so'. Golding (1993: 216), for instance, insists on 'the importance – no, the necessity – to re-cover "urban-ness" in all its anomie, and rather chaotic, heterogeneity, if we are indeed serious about creating a radically pluralistic and democratic society'. Trading on the longstanding radical opposition of the city to the country the emphasis here is on the discrete and fractured qualities of the city as 'the antidote to preserve or maybe invent, against a crushing and totalized "community"...' (p. 217).

Similarly, Iris Marion Young (1990) mobilizes the increasingly mediated, impersonal, and consequently reserved and tolerant qualities of urban social relations first identified by Simmel to give a material location to her poststructurally informed ideal of openness to 'unassimilated otherness'. For Young the city is emancipatory because it promises spaces of relative anonymity, heterogeneity, openness, and change, in which otherness can become unfixed from any totalizing sense of community or self-identity: 'City dwelling situates one's own identity and activity in relation to a horizon of a vast variety of other activity, and the awareness that this unknown, unfamiliar activity, affects the conditions of one's own' (1990: 237–8).

Although Young emphasizes the emancipatory potential of destabilizing identity, her spatial focus on the inner city tends both to fix identity in (inner city) space and to emphasize harmony and stability over other forms of urban experience. While city life as the 'being together of strangers' may condition openness to assimilated otherness, those emancipatory hopes for the being together of strangers are also, as both Nicholas Fyfe in Chapter 3 and Les Back and Michael Keith in Chapter 4 remind us, haunted by the threat of violence and the often repressive measures necessary to deal with it. Indeed Sharon Zukin (1995) argues that fears for personal safety have become increasingly powerful and that this politics of fear is driving the growth of private police forces, gated communities, and public surveillance antithetical to the old civic virtues of civility, security, tact and trust inculcated by mingling with strangers in the city. As such she suggests that the emancipatory city may be a 'residual memory of tolerance and freedom' (Zukin, 1995: 294) that doesn't take into account that the city can be a threatening as well as emancipatory place. Contemporary discussion of the emancipatory city ignores this ambivalence and our responses to it at its peril.

Second, Simmel's account of the increasingly impersonal quality of urban interaction also informs recent utopian hopes for a more democratic and emancipatory public sphere. Richard Sennett, for instance, heralds the reserved and impersonal attitude towards strangers as the foundation for 'civilized existence in which people are comfortable with a diversity of experience, and indeed find nourishment in it' (1974: 340). Like Young, Sennett rejects the communitarian nostalgia for the intimacy of community and face-to-face social relations as a tyranny. While Young sees that nostalgia as simply intolerant, Sennett connects it with a deeper, narcissistic retreat from any concern for distant others and thus to our contemporary politics of personality in which large scale 'forces of domination and inequity remain unchallenged' (1974: 339) by an electoral system driven increasingly by feelings of identification with the character of the candidate rather than impersonal issues about wider social relations. By contrast, Sennett defines civility as 'treating others as though they were strangers and forging a social bond upon that social distance' (p. 264). Drawing on Simmel, Sennett describes how civility is institutionalized in the 'public geography of a city' (p. 264) where strangers meet and interaction is based on social role rather than personal identity. What Sennett celebrates as the impersonal quality of urban life might also be described in Liberal terms as universal insofar as it is based on rules of social interaction derived largely through consensus and applying universally to all citizens without regard to situation or personal identity.

As Gary Bridge discusses at length in Chapter 8, Sennett's understanding of the impersonal quality of urban civility bears important similarities to Jürgen Habermas's (1984, 1987) emancipatory hopes for a public sphere based on ideal speech situations. In an ideal speech situation, claims would be judged based on purely rational criteria about the strength of the argument rather than the identity of the speaker. Thus Habermas, like Sennett, makes the Liberal assumption that the possibility for civilized discourse lies in the elimination of personality and of private interest from civic life and public affairs. Like Sennett as well, Habermas looks to the eighteenth century rise of the modern city to find it in certain privileged urban spaces, like the coffee house, where such speech situations of open and unfettered debate became increasingly possible. This elision of public sphere with public space and the urban raises important questions about whether the 'terms ... are interchangeable, as they often seem to be in the literature' (Mitchell, 1996: 127).

Feminist critics have also questioned the gendered assumptions of this celebration of the public sphere and whether the historic exclusion of women, racial minorities, and others from the urban spaces and practices of open debate was in fact constitutive of them (e.g. Fraser, 1997). In this same vein James DeFilippis and Peter North describe in Chapter 5 how the promise of community participation in the public-private partnership redeveloping the Elephant and Castle in south London was undermined when the state removed the funding for local groups to hire full-time staff with the technical planning expertise and organizing skills necessary to insure their full representation.

Flânerie and the livable city

Perception of the distinct character and quality of modern urban life is closely associated with the unsettling experience of the isolated figure walking alone through the city. It was there in Wordsworth and Simmel and, even more emphatically, in the prose and poetry of Charles Baudelaire. Baudelaire revelled in the 'spree of vitality' he felt amid the hustle of the crowd and the anonymity of the streets (quoted in Williams, 1975: 281). Similar sensations of wonder, excitement, and disorientation while walking the city also feature prominently in the work of Claude Monet, James Joyce, and the film-maker Sergei Eisenstein. The fragmentary and phantasmagoric experience of the city provoked these artists to devise new ways of representing the rhythms of the city such as juxtaposition, montage, stream-of-consciousness, and free association.

Being in the city, Baudelaire wrote, taught the soul to 'give itself utterly, with all its poetry and charity, to the unexpectedly emergent, to the passing unknown' (quoted in Williams, 1975: 281). Contemporary urbanists resort to very similar images of self-abandonment in the face of the spiritually overwhelming 'eroticism of city life, in the broad sense of our attraction to others, the pleasure and excitement of being drawn out of one's secure routine to encounter the novel, the strange, the surprising' (Sandercock, 1998: 210). But as Marshall Berman notes there was a more ambivalent quality to Baudelaire's discussion of the chaos of the streets:

> Baudelaire shows how modern city life forces these moves on everyone; but he shows, too, how in doing this it also paradoxically enforces new modes of freedom. A man who knows how to move in and around and through the traffic can go anywhere ... This mobility opens up a great wealth of new experience and activities for the urban masses. (1982: 159–60)

Walter Benjamin popularized the term flâneur to describe the experience of the stroller and dandy moving anonymously through the crowd and visually consuming the city as a succession of compelling but fleeting impressions. For Benjamin, the reflexive flâneur could grasp the openness and potential transitivity of the city by entering a state of half-wakefulness that, paradoxically, was also one of heightened perception that decentred the subject amid an animated world of objects and produced an experienced, emotional knowledge (*gefühltes Wissen*) (Benjamin, 1983: 525). Similarly as Pinder discusses in Chapter 7, Guy Debord and the Situationists believed that experiments with nomadic movement and playful reappropriation of space offered ways to unsettle the frozen qualities of contemporary urbanism and open up its emancipatory possibilities. By contrast, feminists have responded much more ambivalently to this ideal of the flâneur. For many women, the street is a space of fear rather than of freedom. Feminists like Janet Wolff (1985) have insisted that the flâneur's vaunted freedom and mobility are largely, if not exclusively, the preserve of men of privilege. Other feminists, however, have sought to appropriate the concept of the flâneur to reclaim for women the pleasures and possibilities of the city long denied them (e.g. Wilson, 1992).

This artistic tradition of celebrating the pedestrian's poetic experience of the city finds important contemporary resonance in debates about the liveable city. This discourse of the liveable city was largely inaugurated by Jane Jacobs (1972) and her searing critique of the 'sacking of cities' represented by the proliferation of suburban sprawl and urban renewal schemes. Against the puritanical utopianism of both the suburbanizing Garden City movement (with whose new urbanist descendents Jacobs is now often linked) and Le Corbusierian planning, Jacobs insisted that the city must be diverse, for the juxtaposition of heterogeneous functions, built forms, and people together on the streets was simultaneously the cause and consequence of a thriving public sphere. Making many of the same claims about tolerance in the city made by Simmel and by feminist theorists of identity, Jacobs wrote:

> The tolerance, the room for great differences among neighbours – differences that often go far deeper than differences in colour – which are possible and normal in intensely urban life, but which are so foreign to suburbs and pseudo-suburbs, are possible and normal only when streets of great cities have built in equipment allowing strangers to dwell in peace together on civilized but essentially dignified and reserved terms. (1972: 83)

Her celebration of the city as a liveable, indeed even emancipatory, environment for experiencing difference inspired a generation of inner city gentrification. Like Simmel and Sennett whom she influenced, Jacobs revelled in the dynamic and disorganized but nonetheless orderly ways of being that emerge organically out of the unplanned interactions of strangers on the street. 'City areas with flourishing diversity sprout strange and unpredictable uses and peculiar scenes. But this is not a drawback of diversity. This is the point ... of it ... and one of the missions of cities' (1972: 250–51). The playful disco-socialist activity of the Reclaim the Streets Movement is infected with much the same spirit as both Gavin Brown and Quentin Stevens discuss in their contributions to this volume. The stimulating prospect of this kind of diversity on display is one of the chief cultural attractions of the city for gentrifiers. But there are important issues about both whether their relationship to diversity is engaged and equitable or voyeuristic and appropriative (May, 1996), and about the extent to which in fact the process of gentrification displaces the very sources of diversity and creativity that are the source of its emancipatory promise (Merrifield, 2000).

Conclusion: finding freedom in/and the city

A number of different, and sometimes contradictory, ideas circulate in geography and urban studies under the broad banner of the emancipatory city. It is important to recognize the extent to which contemporary visions of emancipation and the city rehearse older arguments articulated more than a century ago by modernist writers and intellectuals like Marx, Simmel, Baudelaire, and Benjamin. But there are important differences too, which reflect the condition and experience of the postmodern city as the confluence of new global flows and networks.

My review of the genealogy of the emancipatory city raises a number of critical questions that are explored in more detail throughout the chapters that follow. First, descriptions of the city as an enabling, transgressive, or liberating space are based on particular assumptions about liberation, affiliation, and justice, and so we might ask what understandings of freedom and justice does an *emancipatory* city imply? In attempting to answer this question we can draw once again on Simmel. As he explained the freedom promised by the city is:

> not only to be understood in the negative sense as mere freedom of movement and emancipation from prejudices and philistinism. Its essential characteristic is rather to be found in the fact that the particularity and incomparability, which ultimately every person possesses in some way, is actually expressed, giving form to life. That we follow the laws of our inner nature – and this is what freedom is – becomes perceptible and convincing to us and to others only when the expressions of this nature distinguish themselves from others; it is our irreplaceability by others which shows that our mode of existence is not imposed upon us from the outside. (Simmel, 1995: 41)

This distinction between freedom from and freedom to provides a convenient way to classify hopes for an emancipatory city. Thus in describing the effects of controlling the flow of water in the city, Gandy in Chapter 11 emphasizes the negative freedom it has provided from thirst and disease, while Brown in Chapter 6 pays much more attention to the positive freedoms possible in the spaces of the city to perform new forms of identity. But contributors also acknowledge the necessary relationship between the two sorts of freedom. Thus, Fyfe, in Chapter 3, discusses the ambivalent role of CCTV and how its promise of public freedom to enjoy public space free from fear of violence comes at the expense of privacy and freedom from police surveillance. Similarly, both Pinder in Chapter 7 and Stevens in Chapter 9 describe how the creative possibilities for the everyday spaces of the city to be imaginatively refashioned so as to escape from the domination of routine also allow for the expression of new forms of identity and ways of being, which is also a major theme of the other chapters in Part 2 of the book, 'Emancipatory Practices'.

The conceptual task of clarifying these various understandings of emancipation serves not only as one important step in the process both of evaluating their purchase and realizing their potential, but it also provides a way to engage with the ghosts of history that Steve Pile warns us in Chapter 13 haunt the spaces of the city and threaten to interrupt utopian dreams about the emancipatory city.

Second, these various understandings of emancipation raise important questions about tolerance and the relationships between individuals and community in the city. The relationship between tolerance and emancipation is a major theme of Part 1 of the book, 'Cities of (In)Difference'. In different ways all three strands of thinking I discussed in this introduction tended to understand the emancipatory city as a tolerant one and to locate the origins of that ethic in the alienating experience of urbanization. The liberal tradition, in particular, with its strong emphasis on negative freedom from constraint, has tended to understand emancipation as an individualizing experience of urban living freeing the individual from community, conformity, and convention. This liberal suspicion of community as necessarily a form of repressive constraint or special private interest is interrogated in Chapter 4 by Back and Keith and Chapter 5 by DeFilippis and North as well as by Bridge's discussion in Chapter 8 of communicative rationality in the city.

While these chapters, as well as that by Fyfe, follow the sociological tradition of the Chicago School in focusing on the empirically accessible manifestations of tolerance in violence and other behaviour, Chapter 2 by Ruddick picks up on the often neglected psychological legacy of Simmel's (1995) reflections in 'the Metropolis and Mental Life' by considering the psychic dimensions of the encounter with otherness and difference. This theme is also pursued by Pile's emphasis on the unconscious apprehension of the city and by Pratt and San Juan in their discussion of *The Matrix* in Chapter 12. This acknowledgement of the potential fluidity of the subject has important implications for attempts to locate emancipation in the alienating experience of street life, diversity, and difference. If, as Elizabeth Grosz (1992) contends, 'bodies are not culturally pregiven, built environments cannot alienate the very bodies they produce' (p. 249), and so it would follow that 'there is nothing intrinsically alienating' or emancipitory 'about the city' (p. 250). Instead, she suggests that the body and the city are mutually constitutive: 'the city is made and made over into the simulacrum of the body, and the body, in its turn, is transformed, "citified", urbanized as a distinctively metropolitan body' (p. 242).

That understanding of the self and the body as mutually constitutive of the city then raises a third set of questions about where and how those freedoms are to be realized in the city. Discussion of the emancipatory city has tended to focus on the inner city and in its streets, squares, cafés, and other downtown public spaces, the sites through which the strolling flâneur experiences the urban. As such there is a sense in which such accounts may tend to re-enforce the public-private divide. Moreover as globalization, ex-urbanization, and the emergence of a new network society blur longstanding distinctions between the country and the city, there is also the question, posed explicitly both by Ruddick in Chapter 2 and Bridge in Chapter 8, about whether the city can or should lay exclusive claim to the emancipatory qualities of tolerance and cosmopolitan openness to difference. While some critics worry that the narrow focus on the emancipatory potential of certain urban forms amounts to an essentialist spatial fetishism that 'turns a politics of identity into a politics of location' (Bondi, 1993: 98), it might also be asked, more hopefully, whether certain 'urban virtues' might also be found in other kinds of places, sites, and situations. That task of locating an emancipatory city is explicitly taken up in Part 3, 'Utopic Trajectories'. In Chapter 10 Robinson urges us to think about more diverse spatialities of urban transformation, including the imaginary, whose potential is also considered by Pile and by Pratt and San Juan, who insist on the value of ungrounded and unlocatable utopic imaginaries to provide a desired but unattainable horizon of possibility. By contrast Gandy's focus on water provides an important reminder of the underlying biophysical imperatives that any utopian dream must satisfy.

The short reflections on the emancipatory city at the end of the book by Ash Amin and Nigel Thrift in Chapter 14 and David Harvey in Chapter 15 touch on all three of these questions. They, like all the authors in this collection, suggest ways in which we might think more clearly and creatively about an urban politics of emancipation for the future.

Note

1 Of course these and other problems also plague cities in the developing world. It is testimony to the parochialism of urban studies that this official discourse of the liveable city, like the more academic discussion of the emancipatory city in this volume and elsewhere, is largely (but not entirely, see Robinson and Gandy in this volume) restricted to the developed world.

References

Amin, A., Massey, D. and Thrift, N. (2000) *Cities for the Many, Not the Few*. Bristol: Policy Press.

Baeten, G. (2002) (ed.) 'The spaces of utopia and dystopia: introduction', Special issue, *Geografiska Annaler*, 84B: 3–241.

Benjamin, W. (1983) 'Der Flâneur', in W. Benjamin, *Das Passagen-Werk, 2 Bde.* Frankfurt: Suhrkamp. pp. 524–69.

Berman, M. (1982) *All That Is Solid Melts into Air: The Experience of Modernity*. New York: Penguin.

Bondi, L. (1993) 'Locating identity politics', in M. Keith and S. Pile (eds), *Place and the politics of identity*, London: Routledge. pp. 84–101.

Butler, T. and Robson, G. (2001) 'Negotiating the new urban economy – work, home and school: negotiating middle class life in London. Paper presented at the Annual Royal Geographical Society – Institute of British Geographers Conference, Plymouth, 2–5 January.

Chorney, H. (1990) *City of Dreams: Social Theory and the Urban Experience*, Canada: Nelson.

DETR (1999) *Towards an Urban Renaissance: Final Report of the Urban Task Force, chaired by Lord Rogers of Riverside*. Spon/The Stationary Office.

Dowling, R. (1998) 'Neotraditionalism in the suburban landscape: cultural geographies of exclusion in Vancouver, Canada', *Urban Geography*, 19: 105–22.

Ellin, N. (1996) *Postmodern Urbanism*, Oxford: Blackwell.

Fincher, R. and Jacobs, J. (1998) *Cities of Difference*. New York: Guilford Press.

Foucault, M. (1986) 'Of other spaces', *Diacritics*, 16: 22–7.

Fraser, N. (1997) *Justice Interruptus: Critical Reflections on the 'Postsocialist' Condition*. London: Routledge.

Friedan, B. (1963) *The Feminine Mystique*. London: Penguin.

Gold, J. R. (1985) 'The city of the future and the future of the city', in R. King (ed.), *Geographical Futures*. Sheffield: Geographical Association.

Golding, S. (1993) 'Quantum philosophy, impossible geographies, and a few small points about life, liberty and the pursuit of sex (all in the name of democracy)', in M. Keith and S. Pile (eds), *Place and the Politics of Identity*. London: Routledge. pp. 206–19.

Grosz, E. (1992) 'Bodies-cities', in B. Colomina (ed.), *Sexuality and Space*. Princeton, NJ: Princeton Architectural Press. pp. 241–54.

Habermas, J. (1984) *The Theory of Communicative Rationality, Volume 1: Reason and the Rationalisation of Society*, trans. T. McCarthy. London: Heinemann.

Habermas, J. (1987) *The Theory of Communicative Action, Volume 2: Lifeworld and System: A Critique of Functionalist Reason*, trans. T. McCarthy. Cambridge: Polity.

Hanson, S. and Pratt, G. (1995) *Gender, Work, and Space*. New York: Routledge.

Harvey, D. (2000) *Spaces of Hope*. Edinburgh: University of Edinburgh Press.

Heron, L. (1993) (ed) *Streets of Desire: Women's Fictions of the Twentieth Century City*. London: Virago Press.

Jacobs, J. (1972) *The Death and Life of Great American Cities*. Harmondsworth: Penguin.

Keith, M. and Pile, S. (1993) (eds) *Place and the Politics of Identity*. London: Routledge.

Lees, L. (1996) 'In the pursuit of difference: representations of gentrification', *Environment and Planning A*, 28: 453–70.

Lees, L. (2000) 'A reappraisal of gentrification: towards a "geography of gentrification"', *Progress in Human Geography*, 24: 389–408.

Lees, L. (2003) 'Visions of "urban renaissance": the Urban Task Force Report and the Urban White Paper', in R. Imrie and M. Raco (eds), *Urban Renaissance? New Labour, Community and Urban Policy*. Bristol: Policy Press. pp. 61–82.

Lees, L. and Demeritt, D. (1998) 'Envisioning the liveable city: the interplay of "Sin City" and "Sim City" in Vancouver's planning discourse', *Urban Geography*, 19: 332–59.

Marx, K. and Engels, F. (1968) *Selected Works*. Moscow: Progress Publishers.

May, J. (1996) 'Globalization and the politics of place: place and identity in an inner London neighbourhood', *Transactions of the Institute of British Geographers*, 21: 194–215.

McCann, E. (1995) 'Neotraditional developments: the anatomy of a new urban form', *Urban Geography*, 16: 210–33.

Merrifield, A. (2000) 'The dialectics of dystopia: disorder and zero tolerance in the city', *International Journal of Urban and Regional Research*, 24: 473–89.

Merrifield, A. (2002) *Dialectical Urbanism: Social Struggles in the Capitalist City*. New York: Monthly Review Press.

Mitchell, D. (1996) 'Introduction: Public Space and the City', *Urban Geography*, 17: 127–31.

Mumford, L. (1995) 'The culture of cities', in P. Kasinitz (ed.), *Metropolis: Center and Symbol of Our Times*. New York: New York University Press. pp. 21–9.

Park, R. E. (1952) *Human Communities: The City and Human Ecology*. Glencoe, IL: Free Press.

Philo, C. and Wilbert, C. (2000) (eds) *Animal Spaces, Beastly Places: New Geographies of Animal-Human Relations*. London: Routledge.

Pratt, G. and Hanson, S. (1994) 'Geography and the construction of difference', *Gender, Place, and Culture*, 1: 5–30.

Robins, K. (1991) 'Prisoners of the city: whatever could a postmodern city be?', *New Formations*, 15: 1–22.

Sandercock, L. (1998) *Towards Cosmopolis: Planning for Multicultural Cities*. Chichester: John Wiley and Sons.

Sennett, R. (1974) *The Fall of Public Man*. New York: Knopf.

Sennett R. (1989) 'The civitas of seeing', *Places: A Quarterly Journal of Environmental Design*, 5: 82–5.

Short, J. R. (1991) *Imagined Country: Society, Culture, and Environment*. London: Routledge.

Simmel, G. (1995) 'The metropolis and mental life', trans. D. Shill, in P. Kasinitz (ed.), *Metropolis: Center and Symbol of Our Times*. New York: New York University Press, pp. 30–45.

Spigel, L. (1992) 'The suburban home companion: television and the neighbourhood ideal in postwar America', in B. Colomina (ed.), *Sexuality and Space*. Princeton, NJ: Princeton Architectural Press. pp. 185–218.

Stanstell, C. (1987) *City of Women: Sex and Class in New York 1789–1860*. Urbana, Chicago, IL: University of Illinois Press.

UN Centre for Human Settlements (1996) *An Urbanizing World: Global Report on Human Settlements*. Oxford: Oxford University Press.

UN Water World Development Report (2003) *Water for People, Water for Life: Executive Summary of the UN Water World Development Report*. Paris: UNESCO Publishing.

Walkowitz, J. (1992) *City of Dreadful Delight: Narratives of Sexual Danger in Late Victorian London*. Chicago, IL: University of Chicago Press.

Whatmore, S. (2003) *Hybrid Geographies: Natures Cultures Spaces*. London: Sage.

Whyte, W. H. (1956) *Organization Man*. New York: Doubleday.

Williams, R. (1975) *The Country and the City*. St. Albans: Paladin.

Williams, R. (1983) *Keywords: A Vocabulary of Culture and Society*. London: Fontana.

Williams, R. (1989) *Resources of Hope: Culture, Democracy, Socialism*. London: Verso.

Wilson, E. (1991) *The Sphinx in the City: Urban Life, the Control of Disorder, and Women*. London: Verso.

Wilson, E. (1992) 'The invisible flâneur', *New Left Review*, 191: 90–110.

Wilson, E. (2001) *The Contradictions of Culture: Cities, Culture, Women*. London: Sage.

Wirth, L. (1938) 'Urbanism as a way of life', *American Journal of Sociology*, 44: 1–24.

Wolch, J. and Emel, J. (1998) (eds) *Animal Geographies: Place, Politics, and Identity in the Nature-Culture Borderlands*. New York: Verso.

Wolff, J. (1985) 'The invisible flâneuse: women and the literature of modernity', *Theory, Culture, and Society*, 2: 37–46.

Wordsworth, W. (1989) *The Prelude: A Parallel Text*, ed. J. C. Maxwell. London: Penguin. First Published 1850.

Young, I. M. (1990) *Justice and the Politics of Difference*. Princeton, NJ: Princeton University Press.

Zukin, S. (1995) *The Culture of Cities*. Oxford: Blackwell.

As we have seen in the introduction, difference has long been recognized as a feature of urban spaces, and interest in the identity politics of difference has grown. However, in recent years the postmodern turn to celebration of diversity and of the emancipatory potential of difference in the city has come under increasing criticism and, indeed, threat. For example David Harvey, although sympathetic to the politics of difference, triggered a more critical reading of discourses on difference in his *Justice, Nature, and the Geography of Difference*. It is true to say that for every geography of difference that might be detected, mapped, and analysed, there are also present often highly naturalized geographies of *in*difference. The chapters in this section set out to explore and critique ideas of, and responses to, diversity and difference in the city. All three authors point to the domestication of difference in the city, albeit in different ways. They highlight the power relations that both contribute to the constitution of in/difference and interfere in its successful negotiation.

Susan Ruddick warns that the postmodern turn to a celebration of difference has ceded the uninhibited encounter with difference only so much ground and that it may even amount to a sort of colonizing domestication insofar as it tries to make excluded peoples intelligible, comprehensible, and even desirable within new and inclusionary discourses. She is also critical of our seeming inability to rethink the privileged locations of 'the other' beyond the narrow confines of transient public spaces of engagement in the city. Her political hopes are attached to the emancipatory possibilities of 'the monstrous', where, following Donna Harraway the monstrous, that is 'the other', is maintained as a site of radical openness.

Cities have increasingly become sites for sophisticated attempts to regulate diversity. Nicholas Fyfe focuses on so-called zero tolerance policing and the expanding use of closed circuit television surveillance systems (CCTV) in the open/public spaces of the 'indifferent' city. He explores the ambiguity and ambivalence associated with these initiatives and what this means for the 'emancipatory city' thesis. Such initiatives can invade our privacy, making George Orwell's *1984* a reality, but in a different political context they might also be liberating and protective. Fyfe argues that we need to engage with this ambiguity, and he points to work that is looking for new ways to address questions of deviance and diversity in the city outside of the criminal justice system and its agenda of social control.

Les Back and Michael Keith discuss how the relationship between government and urban space is changing in the context of moral concerns over racial tension, community safety, and the behaviour of young people. They see in the subtext of contemporary British social policy an emergent form of self-government that demands particular forms of locality and new forms of selfhood. The city, the ghetto, and even the tower block now serve to organize the public debate and policy practices that structure everyday lives in the multiracial and multiracist city. They argue that it is important to examine how the vocabulary of self-government translates into imagined landscapes of the city that are populated by risks often personified by the young people that are the subject of the research on which the chapter is based. The chapter explores how those micro public spheres sit uncomfortably alongside the civilizing localization of contemporary policy intervention.

James DeFilippis and Peter North critique the political communitarianism that is at the core of an Anglo-American politics that is trying to deal with difference and diversity (in the city). They explore the emancipatory potential of community through a real life study of the conflict that has arisen over the regeneration of the Elephant and Castle neighbourhood of south London. They demonstrate the numerous hurdles that an urban community must overcome in order both to act collectively and to have their voice heard. The chapter offers equal measures of hope and despair. Despite their differences the local community was finally able to act as one voice and to be politically effective but the government's commitment to community-led regeneration and the facilitation of difference in the city seems rather facile in practice, for at the end of the day the council and central government are in control and make the final, rather indifferent, decisions.

2 Domesticating Monsters: cartographies of difference and the emancipatory city

Susan Ruddick

Monsters explained and imagined:

> [A] genus too large to be encapsulated in any conceptual system ... [its] very existance a rebuke to boundary and closure (Cohen, 1996); Grendel the Mettascarppa – the border-stepper in Beowulf; the inhabitants of the New World: the Anthropophagi, the (Headless) Blemmyae; something that induces horror – signifying the uncanny (Schneider, 1999); the monster of Ravenna (borne in Ravenna in 1512 and thought responsible for the defeat of the Italians by the French); the grotesque (Cassuto, 1997), the liminal, any kind of alterity (Cohen, 1996); mutation without tradition or normative precedent (Derrida, 1984), the abject (Shildrick, 2002), King Kong, Hannibal the Cannibal; broken knowledge (Francis Bacon, 1561–1626); prodigies and portents (Ambrose Paré, 1557[1982]) inappropriate/d others (Haraway, 1992); originating from an overabundance of sperm (Aristotle, 384–322 BC), mermaids, the Antipodes (whose feet point backwards/who walk upside down), the slave: 'a special and different kind of humanity' – (Rose, 1976) (approximately 13 million during the 400 years of trans-Atlantic slave trade, currently 27 million world wide); caused by excess sexual delight of the female (Saint Albertus Magnus, 1193–1280); a wonder of nature (Johannes Schenck von Grafenberg, 1530–1598); deformed discourse (Williams, 1996) the monk-calf, the witches of the burning times in Europe (estimated 60–100,000 burned following the Papal bull of 1484); Co-joined twins: Lazarus-Johannes Baptista Colloredo, the hydra, the monsters under your bed, the chimera, the hermaphrodite, the side show freak ...

Monstrous differences and the emancipatory city

I open with this excessive epigraph to make a point: that the course of human history is replete with anxieties about difference, reflected in the range and variety of monsters we have imagined and produced in different historical periods. For many scholars the emancipatory city seems the hopeful locus of a *new way* of dealing with difference – a community 'open to unassimilated otherness' (Harvey, 2000; Jacobs and Fincher, 1998; Young, 1990: 227) and/or the place for a magical urbanism (Davis, 2001). Alternately re-imagined as a city of refuge (Derrida, 2001), the right to the city (Lefebvre, 1968), a new space of social citizenship (Painter and Philo, 1995), cities of difference (Jacobs and Fincher, 1998), citi-state (Pierce et al., 1994). Set sometimes against a failed internationalism, or the betrayal of nation states as guarantors of citizenship, the city is being championed by many as the privileged venue for the emergence of a reformulated human rights. And within the city, it is the intermediate zone of urban experience (Williams, 1989), the open minded public space (Berman, 1986), the new anarchism brought about by city space (Sennett, 1978) that holds its emancipatory possibilities – possibilities which catalyze a re-imagining and re-inventing of connections between those who should be united by their oppressions but are instead divided by their differences.

The project requires a radical openness. But it ventures into difficult and dangerous terrain. It requires that we be able to hold onto that uncomfortable and disquieting moment *before* we collapse the other into someone 'just like us' (the pitfalls of certain forms of class and gender politics) or damn them into an irreconcilable 'them'. It begs the question: 'how are we to handle what is other without robbing it of its otherness?' (Levinson in Mason, 1990: 3).

Several scholars have turned to the monstrous as a strategy for exploring this provocative engagement with difference – as something that *simultaneously* evokes awe, inspiration *even as it horrifies* – a portent that points the way to a new and possible future. It is in this sense that Donna Haraway (1992) writes of the 'promise of monsters'. For Haraway the monster allows the possibility of radical openness with which to engage an in/appropriated other; in/appropriated in the sense that we come to understand the other not on our own terms, but according to a different lexicon. As Haraway notes: 'Articulation *must remain open*, its densities accessible to action and intervention. When the system of connections closes in on itself, when symbolic action becomes perfect, the world is frozen in a dance of death' (Haraway, 1992: 327, emphasis added). And along similar lines Jacques Derrida suggests that:

A future that would not be monstrous would not be a future; it would already be a predictable, calculable, and programmable tomorrow. All experience open to the future is prepared or prepares itself to welcome the monstrous arrivant, to welcome it, that is, to accord hospitality to that which is absolutely foreign or strange, *but also, one must add, to try to domesticate it, that is, to make it part of the household and have it assume the habits, to make us assume new habits.* This is the movement of culture. (1995: 387, emphasis added)

What are we to make of this project? Is it possible to imagine a transformative politics, which rests on a new ability – not to dissolve difference into a 'reformulated universality,' but to sustain 'multiple debinarized, fluid ever-shifting differences' (Fraser, 1995; Jacobs and Fincher, 1998)? And if so, what are the implications of privileging the city and the public spaces within it as the catalytic venues for this new venture? The political challenge of embracing difference only truly arises at the moment of our profound uncertainty: that point at which this difference threatens to undermine, destabilize, or re-organize our most cherished beliefs, when we stand on the problematic terrain where the status of the other still hangs in the balance.

In the past few decades, border theorists have stressed the hopeful possibilities of redefining these borders, not as definitive edges but fertile zones where new kinds of hybridities might emerge (Anzaldua, 1987). The persistence of these zones, however, depends not on a new set of combinatory rules to 'create' border zones but an abandonment of the strictures that maintain binaries. And this in turn, requires a deeper understanding of how we are schooled in identifying and locating difference. There are powerful dynamics at work that threaten this hopeful celebration of the monstrous, that prevent it from even coming into being. Before we can embark on such a project, therefore, I would argue that we must explore more negative representations that undermine this project, that make it even un-imaginable.

In this chapter then I will look at the discourses of the monster that help to order, contain and control our experience of difference in the city – the ways in a sense we 'domesticate monsters' – 'adapting them for life in intimate association with man' (*Webster's New Collegiate Dictionary*, 1976: 339) through narrative accounts that help us to contain or collapse difference. But I am also intrigued with the possibilities of monsters domesticating us – hence the ambivalence of the gerund in my title 'Domesticating Monsters' – the transformative possibilities that monsters might provide us.

Exploring *the idea of monsters,* rather than focusing on any one particular 'other', allows us to come to grips with the ways that the 'dark side' o difference *prefigures* our imagination and stalks the horizons of our consciousness. If we hold the city as the locus for a re-imagining of our relationship to difference, then it becomes important to understand first *how* difference becomes readable in these antipodal forms – monster/horror versus monster/wonder. What is the taxonomy of the monstrous, broadly speaking? In one sense this is an impossible question – for the defining quality of the monstrous is that it *defies* taxonomy, and yet our stories of monsters often share common qualities (as I will argue in the following pages) – although their capacity to horrify lies in the novelty of their particularties. What are the structuring taxonomies that enable the monstrous to become animated in a variety of ways – in this case the specificities of excess and lack, of visibility and invisibility, that make these monsters recognizable?

And secondly and equally important it is necessary to explore *where* different versions of the monster become authorized – what is the *syntax* of the monstrous, or more properly the politics of location that gives it resonance? Why, for instance, do we privilege the celebration of difference in certain places while expelling it from others? And what does this politics of containment mean for the larger project of the emancipatory city?

We must therefore also investigate its geography. What are the multiplicity of sites that contain or displace the monstrous – where is monstrosity performed? What are its strategies and tactics of mobility – what permanence or lack thereof is it accorded in the city? It is through this investigation hopefully that we can begin to understand how the monstrous plays in and around, and often against, the cartography of the emancipatory city. Of course this is potentially an enormous undertaking: the sites where we might encounter the monstrous are infinite; as Shildrick (2002: ix) notes, 'the monstrous is not only an exteriority', and thus 'we are always and everywhere vulnerable' to its disruptive potential. Nevertheless each historical moment evokes a particular set of contested terrains, particular sites where the bounded-ness of meaning, the sense of what is right and proper, is troubled. In this chapter, drawing largely on representations in film, I would like to focus on two dominant and antipodal narrative themes about monstrosity as they relate to the city and to our in/ability to welcome difference – these are stories about monsters who are visible, immediately legible and apparently ex-centric in their origin (the Godzillas, King Kongs, and Frankensteins of the world), and stories about monsters who are invisible, not recognized on sight and apparently intrinsic in their origins, (in this case, the Hannibal Lectors, the Dorian Grays, the Dr Jekyls and Mr Hydes). These stories moreover, gesture to the range of ways that difference is constituted – difference as a kind of radical alterity – unknown and unknowable, wholly 'other'; difference as a kind of associative alterity – subject to gradation and comparison;[1] difference in other peoples and social groups, and difference contained within ourselves (see Levinas, 1998; Mason, 1990).

We are, however, not strangers to difference, nor to difference that is constituted as monstrous. The image[2] in Figure 2.1 of Godzilla and Fay Rae intrigues me because it speaks volumes to the ways in which the iconography of 'the monstrous' provides us with a ready made geography for both recognizing and locating the unimaginable and horrific. With its subtitle 'September 11: A sleeping giant has been awakened', it is the most graphic example of the resurrection of the King Kong story in American media as a way to depict and contain the horror following the destruction of the World Trade Center. And it suggests that even as we experience 'the monstrous' in its unmediated form, truly, in the horrors of war, in the night terrors of childhood, or alternately on other occasions, in the mis-recognition of difference, that we are already well-schooled in its politics of location, through myth, through stories, games and other means. We may be uncertain in the initial days and months following an event or encounter where and how to locate it – in fact

September 11, 2001: A sleeping giant has been awakened.

FIGURE 2.1 September 11, 2001: A sleeping giant has been awakened

it is this unsettling ambivalence, this 'refusal to be properly located' that gives 'the monstrous' its power (see Schneider, 1999).

But eventually the image settles. And this settling begs the question: what are the coordinates of this politics of location? How do we come to know 'where' to place our anxieties?

Children, arguably, first experience the monstrous as part of a 'deep interior' – the 'monsters under the bed' that surface in the moments between sleep and wakefulness, that terrorize if children (or any of their body parts) stray from the sanctity of the covers. But they learn over time to treat these fears as groundless. Part of this process is achieved through normalizing these monsters in children's literature and games: *Where the Wild Things Are, The Monster at the End of this Book,* or most recently *Monsters Inc.* teach children that difference is of no consequence – that others are 'just like us' underneath all that green fur. Alternately video and computer games involving the destruction of monsters teach one that monsters are: a) easily differentiated from us, and b) must be destroyed. These particular mappings set us up for two clearly defined trajectories in dealing with difference: difference is alternately collapsed or contained. And more importantly significant differences are always represented as if they lie outside the bounds, or trouble the edges, of sanctioned culture, rather than emanating from within.

Monstrous ex-centricities

The first set of stories I would like to explore address the 'monster from without' – 'monsters against the city', monsters run amok. Although the form that these monsters take varies, the narrative is familiar. The story goes like this: A monster (usually as a singularity, not a race or species) threatens the city/civilization by raiding or destroying its civic spaces, by exacting a sacrifice from the inhabitants, by constricting freedom of movement beyond its boundaries. For order to be restored the monster is destroyed. The story of King Kong and the many stories of Godzilla, even Frankenstein, come to mind but they are not without precedent, – they are traceable back to the gleeful rampages of Grendel in Beowulf,[3] and they share a common structure in terms of the attributes signaling their difference, the nature of their alterity, and their location of origin.

The first noteworthy taxonomic feature of these particular monsters (although there are others, which I will deal with later) is that they are clearly recognizable as such – knowable on sight because their bodies signify inexplicable excess, and more importantly because the cultural labour that positions them as such precedes our first encounter with them.

And a second feature common to these tales and pertaining to their geography is that these monsters are apparently eccentric – not simply outside the centre, or organized in subordinate relation to the centre. We are taught to think of them as appearing both literally and figuratively from beyond the boundaries of the known world. They come from an earlier time, from the depths of the sea, or cower in the attics and cellars of our consciousness, under beds, in sewers, emerging at the hour between wakefulness and sleep. They gesture to a kind of radical alterity, an unknowable difference – one which we cannot experience.

But significantly the initial contact with the monster originates as a result of a human generated disturbance in the form of excess: humans venturing too far beyond the boundaries of their known world (Godzilla), or by exceeding the ethical boundaries of science (Frankenstein) – thus in a sense producing or awakening the monster. This sense that their mobility is somehow catalyzed by human intervention speaks to the trace of the monster in the interiority of our culture – the sense that they are never wholly outside but trouble the borders of the civilized world in part through human desire or failing. Thus, they are not wholly ex-centric: we are complicit in some way in their appearance.

Godzilla, in the original story for instance, is discovered by a group of Japanese fishing boats in the region of the Ohto Islands. And the vengeance he/she/it? wreaks on Tokyo in successive films becomes a metaphor for a variety of contemporary processes – first the destruction of Hiroshima and Nagasaki by American atom bombs, later the too rapid modernization of Japan, and a call to return to a more ancient way of life. King Kong is also to some extent a product of hubris – the audacity of an arrogant filmmaker and adventurer who sets out to film the wild (a theme re-iterated later in the movie *Jurassic Park*).

These stories teach particular lessons about dealing with difference: the combinatory aspects of danger and desire that warn us against any form of contact. And the threat is a threat not only to us personally – or to those engaging in these foolish experiments (the story goes) but a threat to our publicity, our civility, and our spaces of civility. Through contact with the 'outside' we bring these monsters into the centres of our worlds. The destruction of public spaces of the city, the havoc wreaked on innocent unknowing civilians is a central and enduring theme in these tales. This is the strategic relation of this form of the monstrous to our city spaces – bringing unthinking destruction to our spaces of public civility.

The appropriate response, however, cannot be simply to treat these monsters as portent or wonder. In practice, even when we let difference into the public spaces of the city in this way, its relation tends to be a tactical one (in the sense introduced by de Certeau, 1984), an action that plays on and within a power organized by another. Peter Goheen's (1993) study of the street of the nineteenth century, for instance, suggests that at carnival time, the street became a place for the temporary celebration of excess. Public space is afforded the role of chameleon, a temporary transformation. Excess is performed there, but it is also contained.

The corporeality of these monsters also intrigues me, as they are simultaneously positioned as excess and lack. In these first set of stories the combination of excess and lack marks the bodies of monsters in particular and legible ways. King Kong's story of fascination with Fay Rae, for instance, becomes troublingly impossible in part because of his sheer size (and curiously he grows in size when displaced from his 'natural' habitat to New York) – a human sized King Kong would have evoked a story about the possibilities/impossibilities of domestication rather than the inevitabilities of destruction – the great Kong as (failed) pet – a latter day Sasquatch or 'Big Foot' rather than monster. But the combinatory dynamics of excess/lack become mobilized in a variety of ways when traced across the bodies of these beings made monstrous. Not only do they effectively displace them from the spaces of civility that they invade, *they split them from their places of origin.* Godzilla and King Kong for instance function as aberrations – certainly not belonging to the spaces they invade but also not 'really' belonging to the places from which they emanate. Godzilla is seen as a monster from another time, Kong is unique, an anomaly, simultaneously produced by nature and at the same time un-natural.

This practice has often been a tactic of colonization in the narratives we create about other peoples. Thus Peter Mason (1990) notes, the emphasis of sixteenth-century explorers of the New World on the nudity of the natives, their representation as sexual excess, the emphasis on their apparent cruelty, their apparent irrationality, the denial of their long history as agriculturalists (in fact the Algonkin grew enough food to feed both themselves and the English) and the falsification of their history as hunter gatherers (Duverger, 1983: 226–7) created, through a violent rift, the conceptual space for colonization. As Mason puts it so well – in the eyes of the colonizers:

> Topographically, the land may not seem so strange, but the behaviour of the inhabitants with respect to their land is somewhat puzzling. *This figure has a strong rhetorical force that justifies intervention to repair the rift and to restore harmony – if the natives cannot do the job properly, they will have to let others do it for them.* (1990: 178, emphasis added)

This splitting of a people from their land through representation as excess and lack has been repeated in successive histories of invasion, colonization and internal subordination and repression of others – represented as 'almost by nature' excessive zealous, lacking rationality, a throwback to an earlier era. I would argue that the subordination of immigrant peoples *within* our cities also invokes this split from their lands of origin – a kind of doubled politics of dislocation, achieved through the representation of their relation to *both* places as excess and lack, a representation traced out and 'knowable' through their bodies.

FIGURE 2.2 The Irish Frankenstein

One can see similar strategies at work in the representations of peoples 'at home and abroad' – albeit with different coordinates. Seen in Figure 2.2 for example, England's colonization of Ireland is justified through a 1882 cartoon in *Punch*, a British political magazine, depicting the Irish as a kind of Frankenstein. But we find similar strategies in pre-Holocaust representations of Jews in Europe (as intellectual excess incapable of controlling their greed), blacks in Africa (as sexual excess, incapable of managing their resources), and more recently Arabs in the Middle East (as religious zealots) belying more complicated historical geographies and more difficult complicities between the 'West' and the 'Other'.

The intrinsic monster; monster as false civility

Against these stories of immediately legible and ex-centric monsters we have the monster as mask – a kind of false civility concealing a deep perversion. In this version of the monstrous, the monster is not immediately legible. Contemporary films such as *Silence of the Lambs* and *Hannibal* are exemplary of this form, but we might also consider many others – older stories such as *Dr Jekyl and Mr Hyde* or *The Picture of Dorian Gray*, or more recently *Psycho*, *A Clockwork Orange* and *The Pledge* come to mind. An antipodal form to the apparent *eccentricity* and *legibility* of King Kong or other such monsters, these monsters on first encounter could be anyone, look like anyone, live anywhere. In fact they don't just look like 'anyone': they often appear as icons of civility – bourgeois white males in positions of power or prestige. But these are not stories critiquing elites. Rather these individuals are meant to represent 'everyman'. Their transgressions speak to the fears and desires of a wider populace: fear of aging (*Dorian Gray*), fear of the evil sides of ones character (*Dr Jekyl and Mr Hyde*), a sense of absolute power (*Hannibal*). But in all of these stories, taken to their extremes, these individuals become monstrous – they are revealed as monstrous through their actions, but actions made known in private through intimate association or extended contact.

These monsters are marked by a taxonomy that 'splits the difference' – the difference within ourselves – often speaking to the dangers (the story goes) of not properly integrating our darker sides. Thus Dorian Gray relegates his aging self to a portrait in his attic; Dr Jekyl the nastier sides of his character to his alter ego, Mr Hyde; Hannibal Lector a psychiatrist whose flirtation with unfettered freedom from the bonds of civility takes the form of cannibalism. These monsters are not recognized on sight, although we have since come to recognize them. The face of Anthony Hopkins as Hannibal Lector is by now almost irretrievably linked with the character, a white haired somewhat ordinary looking older man who comes to stand for cannibalism. But the heightened shock of recognition comes from the fact that this monster emanates from the centre of bourgeois male civility. Hannibal would have played very different as a black man or a white woman.

The peculiar taxonomy of these monsters is then characterized by an invisibility – an absence of recognizable monstrosity – it lies rather in the juxtaposition of apparent normalcy and concealed perversity – individuals who seem like us in every respect, yet are capable of unspeakable acts.

And the taxonomy of invisibility is intimately linked to the cartography of these monsters – where their monstrosity is in a sense performed. Apparently normal in public where initial encounters often occur, their monstrosity is revealed in private. They do not run amok through city centres, they pass un-noticed, appear as any man, every man. There is no a priori cultural labour to signal their arrival. In contrast to the ex-centric monsters we encountered earlier, these monsters emanate from within our borders – they originate not outside of the city or civilization but outside of logos, outside rationality. In spatial terms then, they are apparently intrinsic rather than ex-centric. Moreover, they haunt and in-habit the private spaces of the city – its houses, churches, schools, prisons, and hospitals.

In these stories the monster is linked to our spaces of vulnerability, privacy, but also communality – the familiarity and intimacy of the everyday. And so everyday objects become imbued with a sense of horror and danger – the silhouette of a woman projected on a shower curtain immediately recalls the scene from *Psycho*; a gourmet meal, the *Silence of the Lambs*.

Moreover, these monsters are not treated as a singularity – an anomalous occurrence, but rather signal the possibility for transgression in everyone – a kind of associative alterity which addresses issues with which we all must struggle. Thus for instance in the film *Hannibal*, the contrast of Lector and Clarice the police detective, suggests that she struggles with similar issues around freedom and possibility, in the constricting nature of her job as a female detective in a largely male profession. The issues these monsters struggle with are those common to us all – their monstrosity emerges from an inability to patrol cultural boundaries and maintain cultural norms. This masked form of monstrosity is not a new feature of the urban landscape. Lamenting the loss of integrity that he felt symptomatic of the modernizing city, Rousseau (1758 trans., 1968) wrote:

In a big city full of scheming idle people without religion or principle, whose imagination depraved by sloth, inactivity, the love of pleasure, and great needs, engenders only monsters and inspires only crimes; in a big city, *moeurs* and honour are nothing because each, easily hiding his conduct from the public eye, shows himself only by his reputation. (cited in Sennett, 1978: 118)

But anxieties about the mask are perhaps equally triggered by a different set of fears – that of marginalized peoples who are often forced to 'wear the mask' to downplay difference in order to 'fit in' – fears around those who 'pass' in order to gain a passport to another life, fears about those whose apparent complicity belies a greater knowledge of the systemic inequities that limit their actions.

Son of former slaves and the first nationally recognized American black poet and novelist Paul Laurence Dunbar wrote in 1896 a poem describing the doubled position of African Americans in that period:

We Wear the Mask

We wear the mask that grins and lies,
It hides our cheeks and shades our eyes,
This debt we pay to human guile;
With torn and bleeding hearts we smile,
And mouth with myriad subtleties.

Why should the world be over-wise,
In counting all our tears and sighs?
Nay, let them only see us, while
We wear the mask.

We smile, but, O great Christ, our cries
To thee from tortured souls arise.
We sing, but oh the clay is vile
Beneath our feet, and long the mile;
But let the world dream otherwise,
We wear the mask!

(From *We Wear the Mask*, Paul Laurence Dunbar, 1993)

And bell hooks (1989) writing almost a century later expresses a similar sentiment when she describes the surprise that white feminists feel when white women are characterized by black women as parasitic, as living off their (domestic) labour. The fear of masked civility here is the fear that others might be fully aware of the conditions of their oppression. And cognizant and critical of their oppressors. These stories suggest an alternative reading of the spatiality of difference in the city as well – it is not simply the masked civility of public spaces that must be re-negotiated in a recognition of difference, but the unknown or intimate spaces of the city – those cleaned by janitors, cared for by 'hired help', tended by maids, watched over by nannies, to name but a few. The film *The Long Walk Home* about the 1955 Montgomery bus boy-cott charts the shifting relationship of a white family to their black maid, who has decided to participate in the civil rights movement. As the story unfolds the actions of the maid split family loyalties right down the middle as her refusal to take the bus to the house where she works, is secretly supported by the wife of the house and adamantly opposed by her (Klan member) husband.

Even the most highly segregated or partitioned economic or ethno-cultural spaces are already and always permeable to difference. Our relationships with difference cannot be neatly contained but are knotted in an intricate tangle of public and private spaces and relationships in the city.

Beyond impossible divisions …

How then do we address these impulses to contain and control difference? One response is to resurrect the division between monster as horror and monster as portent and simply emphasize the latter – arguably this tendency can be found to some degree in new writings on magical urbanism which seek to challenge stereotypical views of immigrants by celebrating the dynamism and creativity of Latino/a culture (Davis, 2001) – a kind of celebratory gesture towards difference. I have no quarrel with these writings in and of themselves, but we must take care to push the boundaries of these kinds of representations, or they may end up providing a limited space for the celebration of the exotic – in the sense that Roland Barthes writes about – merely existing in a space already prepared for difference (Barthes, 1983).

To treat those different from us – as 'signs taken for wonders' is no guarantee of open-mindedness to difference: in fact it can produce its opposite. We still celebrate Greek and Roman cities as crucibles for democracy, taking the agora as the starting point for the emergence of a polity. There is a kind of wistfulness about writings celebrating the agora as if a kernel of 'true democracy' might be nurtured within this city space and gradually spread outward to encompass humanity. But this same period produced Pliny the Elders writings on the imaginary races of people that came to be known as the Plinian races (see Figure 2.3). These included such fantastic peoples as the *Astomi* who have no mouth but live off the aroma of fruits; the *Blemmyae* who have no head but whose faces appear on their chest. Many of the later writings on the much celebrated Plinian races focus on the ways that these imagined peoples served to link difference to distance – to suggest as Sibley notes that 'the ancient Greeks and Romans saw themselves at the center of the civilized world and in their ordering of cultures and societies, the farther a group was from the center the greater was its vice' (Sibley, 1995: 50; see also Mason, 1990). One line of thinking about this is to suggest that difference and distance are intimately related and that through prolonged contact notions about the absolute strangeness of others diminish.

But these interpretations of Pliny overlook the ways that 'difference as distance' also serves to constitute divisions *within the interiority* of cultures. And they sometimes allow Pliny's writings to be treated as fantastical – the Plinian races as 'marvel at the margins'. In a re-reading of the Plinian races through the multiply constituted poles of the other we might trace an alternate

FIGURE 2.3 The Plinian races

connection. In a society where it was possible to feed slaves only corn, why would it not be possible to imagine a race of men who lived off the aroma of apples (the Astomi)? And in a society that positioned women as chattel, would it not be possible to imagine a race of other peoples distinguished by an excess of orifices? In such a reading the Plinian races come to serve a very different purpose: a naturalizing function for an existing social order, not a liberating infusion of the fabulous. The geography of these visible recognizable monsters – whether Godzilla, King Kong or the Plinian races – thus both *exceeds* the spatio-temporal boundaries of the city and has been lodged and contained *deeply within its interior*.

This omission is significant because it allows us simultaneously to read the monstrous races as a fabulous tale, and on the other hand it allows us to elevate the Greek agora and Roman fora to the status of iconic civility, with the assumption that the deeply patriarchal and slave-based foundations of these societies were somehow incidental, or could be stripped away, ignored, forgotten, to protect the kernel of a democratic ideal that the agora enshrined. And it raises dangers about thinking that might suggest we could build an emancipatory city without simultaneously considering what happens beyond its borders.

... and dangerous fusions

In the current conjuncture moreover, for all our hopefulness about the progressive potential of cities, we are on the verge of a dangerous fusion which threatens to combine, in our worst nightmares, the stories of monster as ex-centric and monster as intrinsic. This fusion has happened at different points in history to produce a kind of radical alterity – a position of condemnation from which there is no escape, and which provides justification for an infinite variety of persecutions; where those so positioned can neither rest comfortably in an assignation as 'other' nor be recognized as 'just like us', but become suspect in either positioning. In the recent past this positioning was marked by the young black man who was damned if he dressed differently or sported what appeared to be gang attire but also held in suspicion if he looked normal or drove a decent car (an up and coming drug dealer? a stolen car? – ill-gotten gains, was the line of thinking). Or the so-called witches of the fifteenth century who could 'pass' for human but then take on different forms and whose only proof of innocence was death.[4] The current candidates for this positiing as radical other are young Muslims who when wearing Kaffiyehs or turbans are viewed suspiciously as part of a backward people, but when they adopt modern dress are equally suspected of being 'sleeper terrorists'. In both cases these narratives about monsters – one ex-centric and legible, the other intrinsic and un-recognizable – function as gatekeepers that prevent us from examining productively and openly potential sites of transgression. They prevent us from re-imagining others and ourselves by marking any enterprise of encounter as inherently dangerous.

Why does this matter to the project of the emancipatory city? Why do we need to think of these kinds of difference and the spaces that engender them together rather than separately – both within and beyond the city limits?

Why does the mixing of discourses and bodies become linked and limited to transient fleeting encounters in public spaces – acts of safe consumption in ethnic restaurants or neighborhoods or momentary performances in carnivals? Why not the re-imagining of home space, of institutional spaces, workplaces, social services, ideas of citizenship? The discussion is fixated on a kind of compulsory cosmopolitanism that ignores the way boundaries have been redrawn. In our contemporary fixation on public spaces we have lost the ability to re-imagine the prison, the mall, the gated community, the hospital, the school – the infinite variety of places that might be dismatled or refashioned as truly welcoming.

It is as if we have ceded an uninhibited encounter with difference only so much ground – a temporary, transient public space of engagement which may, for a brief time, disorient or arouse the participant but which leaves the structures of logos firmly intact – and as well a space for retreat from the vertigo induced by fleeting encounters with the other. For all its gesture towards a celebratory kind of difference these discourses of the emancipatory city occlude the places where the monstrous is maintained as a horror. It would be tempting at this point to move to an analysis of more open-minded representations of difference. In fact several films come to mind: *The Long Way Home*, mentioned earlier; *Antonia's Line*, a Dutch film about a loosely matriarchal commune established on the outskirts of a Dutch village after the Second World War; or *The Theory of Flight* a story about disability, sexuality and the relationship between a young woman dying of motor neuron disease and her caregiver. And while these would provide examples of transformative encounters with difference they would not necessarily open us to the process of embracing difference but merely sanction 'established others' that are now worthy of this encounter. As Derrida (1989: 80) writes so pointedly 'our monsters cannot be announced. One cannot say "Here are our monsters" without immediately turning the monsters into pets' (without, it is implied, domesticating the very thing we are trying to remain open to).

The success of the city as an emancipatory project depends critically on our ability to rethink the privileged locations of 'the other' beyond the narrow confines of the street, beyond the cacophony of particular neighborhoods – into the more intimate spaces of the 'unknown city' and beyond the putative boundaries of the city to other places. And at the same time it requires us to think *the connectivity between* – or perhaps put more strongly the *inseparability of* – what we have come to think of as different kinds of othering: the other in ourselves; the other in different people as individuals (men and women, abled and dis/differently abled, straight and queer); and the other in peoples of different social, class-based and ethno-cultural groups.

If we really believe in the emancipatory possibilities of the monstrous and the project of truly welcoming difference, then we need to be prepared to accept it into the spaces where it is likely to domesticate us, and these are as often as not, *not* the (limited public) spaces we have already prepared in our imaginations.

Notes

1 Although the differences between radical and associative alterity are clear in their conception they become muddier in their application. Although in this text I have tended to emphasize radical alterity with the first set of stories and associative alterity with the second it probably makes more sense to say that both share elements of each kind of alterity.

2 Following the destruction of the World Trade Center on September 11 2001, the American media made numerous references to King Kong as a way of representing the monstrous attack on New York. This image appeared on a website for public distribution. See http://www.filmlandclassics.com/dir/news/sleeping_giant.html

3 But for an interesting alternative, although distinctively nihilist take, read Gardner (1989).

4 We can extend this thinking to the impossible position that Jews faced in Europe during the pogroms, or blacks in the slave trade.

References

Anzaldua, G. (1987) *Borderlands/La Frontera*. San Francisco, CA: Aunt Lute Books.

Barthes, R. (1983) *Mythologies*, trans. Jonathan Cape. New York: Hill and Wang.

Berman, M. (1986) 'Taking it to the streets: conflict and community in public space', *Dissent*, Summer, 33(4): 476–85.

Cassuto, L. (1997) *The Inhuman Race: The Racial Grotesque in American Literature and Culture*. New York: Columbia University Press.

Cohen, J. (1996) 'Monster culture (seven theses) in monster theory', in J. Cohen (ed.), *Monster Theory: Reading Culture*. Minneapolis, MN: University of Minnesota Press. pp. 3–25.

Davis, M. (2001) *Magical Urbanism: Latinos Re-Invent the US Big City*. New York: Verso.

de Certeau, M. (1984) *The Practice of Everyday Life*, trans. S. Randall. Berkeley, CA: University of California Press.

Derrida, J. (1984) 'Deconstruction and the other', in R. Kearney (ed.), *Dialogues with Contemporary Continental Thinkers*. Manchester: Manchester University Press. p. 123.

Derrida, J. (1989) 'Some statements and truisms about neologisms, newisms, positisms, parasitisms, and other small seismisms', in D. Carroll (ed.), *The States of Theory*, New York: Columbia University Press. pp. 63–94.

Derrida, J. (1995) 'From traumatism to promise', in E. Weber (ed.), *POINTS – INTERVIEWS, 1974–1994*. Stanford, CA: Stanford University Press. pp. 385–7.

Derrida, J. (2001) *On Cosmopolitanism and Forgiveness*. London: Routledge.

Dunbar, P. L. (1993) 'We wear the mask', in J. M. Braxton (ed.), *The Collected Poetry of Paul Laurence Dunbar*. Charlottesville, VA: University Press of Virginia. p. 71.

Duverger, C. (1983) *L'Origine des Azteques*. Paris: Editions Seuil.

Fraser, N. (1995) 'Recognition or redistribution? a critical reading of Iris Young's Justice and the Politics of Difference', *Journal of Political Philosophy*, 3: 166–80.

Gardner, J. (1989) *Grendel*. New York: Vintage Books.

Goheen, P. G. (1993) 'The ritual of the streets in mid-nineteenth century Toronto', *Environment and Planning D: Society and Space*, 11: 127–46.

Haraway, D. (1992) 'The promise of monsters: a regenerative politics for inappropriate/d Others' in L. Grossberg, C. Nelson, and B. Treichler (eds), *Cultural Studies*. New York: Routledge. pp. 295–337.

Harvey, D. (2000) *Spaces of Hope*. Edinburgh: Edinburgh University Press.

hooks, b. (1989) *Talking Back: Thinking Feminist. Thinking Black*. Toronto: Between the Lines.

Jacobs, J. and Fincher, R. (1998) 'Introduction', in R. Fincher and J. Jacobs (eds), *Cities of Difference*. New York: Guilford Press, pp. 1–25.

Lefebvre, H. (1968) *Le Droit à la Ville*. Paris: Anthropos.

Levinas, E. (1998) *Entre-Nous*, trans. M. B. Smith and B. Harshav. New York: Columbia University Press.

Mason, P. (1990) *Deconstructing America: Representations of the Other*. London: Routledge.

Painter, J. and Philo, C. (1995) 'Spaces of citizenship: an introduction', *Political Geography*, 14: 107–20.

Paré, A. (1982) *On Monsters and Marvels*, trans. and intro. by J. L. Pallister. Chicago, IL: University of Chicago Press.

Pierce, N., Johnson, C. and Hall, J. (1994) *Citistates: How Urban America Can Prosper in a Competitive World*. Washington, DC: Seven Locks Press.

Rose, W. (1976) *A Documentary History of Slavery in North America*. Oxford: Oxford University Press.

Rousseau, J. J. (1968) *Politics and the Arts: The Letter to M. d'Alembert*, trans. A. Bloom. Ithaca, NY: Cornell University (orig. publication 1758).

Schneider, S. (1999) 'Monsters as (uncanny) metaphors: Freud, Lakoff, and the representation of monstrosity in cinematic horror', *Other Voices*, (3): January.

Sennett, R. (1978) *The Fall of Public Man*. New York: Vintage Books.

Shildrick, M. (2002) *Embodying the Monster: Encounters with the Vulnerable Self*. Thousand Oaks, CA: Sage.

Sibley, D. (1995) *Geographies of Exclusion*. London: Routledge.

Webster's New Collegiate Dictionary (1976) Toronto: Thompson Allen and Sons.

Williams, D. (1996) *Deformed Discourse: The Function of the Monster in Mediaeval Thought and Literature*. Montreal: McGill-Queens University Press.

Williams, R. (1989) *The Politics of Modernism*. London: Verso.

Young, I. M. (1990) *Justice and the Politics of Difference*. Princeton, NJ: Princeton University Press.

3 Zero Tolerance, Maximum Surveillance? deviance, difference and crime control in the late modern city

Nicholas Fyfe

Two journeys in New York City

Journey 1: In *The Conscience of the Eye*, the urban social theorist Richard Sennett guides us on a three mile walk from his apartment in Greenwich Village. The walk leads first through a 'drug preserve' just east of Washington Square, where addict-dealers in cocaine are never still, 'their arms jerky, they pace and pace; in their electric nervousness, they radiate more danger than the old stoned men' who used to occupy this area (1990: 124). Sennett then heads along Third Avenue to the edges of the Grammercy Park district ('the people who live here are buyers for department stores, women who began in New York as secretaries'); and then into the middle Twenties between Third and Lexington (where there are 'bars that cater to ... leather fetishists, bars in run-down townhouses with no signs and blacked out windows' (1990: 129). In the upper Twenties along Lexington Avenue, Sennett observes the 'bags of spice [that] lie in ranks within the shops run by Indians and Pakistanis'; and then the 'last lap' of his walk passes through Murray Hill ('a quarter of the old elite in New York') and on to the east Forties between Lexington and First avenues from where one can see miles of burned out buildings and bricked up windows. 'This permissible belt of desolation in so rich a city', Sennett (1990: 131) concludes, 'is like a boast of civic indifference.'

Journey 2: The second journey is an account by William Bratton of arriving in New York City in 1990 as the new Chief of Police of the City's Transit Police Department:

I remember driving from LaGuardia Airport down the highway into Manhattan. Graffiti, burned out cars and trash seemed to be everywhere. It looked like something out of a futuristic movie. Then as you entered Manhattan, you met the unofficial greeter for the City of New York, the Squeegee pest. Welcome to

New York City. This guy had a dirty rag or squeegee and would wash your window with some dirty liquid and ask for or demand money. Proceeding down Fifth Avenue ... unlicensed street peddlers and beggars were everywhere. Then down into the subway where everyday 200,000 fare evaders jumped over turnstiles while shakedown artists vandalised turnstiles and demanded that paying passengers handed over their tokens to them. Beggars were on every train. Every platform seemed to have a cardboard city where the homeless had taken up residence. This was a city that had stopped caring about itself. (Bratton, 1997: 33–4)

Despite their contrasting backgrounds, Sennett and Bratton's accounts of difference and deviance in 1990s New York City contain some intriguing similarities. Both are deeply troubled by the sense of indifference that characterizes the city. Sennett's concerns stem from the way his experiences of walking through a city seem so unlike that of the Parisian poet and flâneur, Baudelaire. The latter appeared engaged with the world around him, yet Sennett finds himself, as well as the others he observes, simply indifferent to each other:

New York should be the ideal city of exposure to the outside ... By walking in the middle of New York one is immersed in the differences of this most diverse of cities, but ... [a] walk in New York reveals ... that difference from and indifference to others are a related and unhappy pair. The eye sees differences to which it reacts with indifference.' (1990: 128–9)

For Sennett, however, there are emancipatory possibilities in this city of indifferent strangers. Developing an argument concerning the need for heterogeneity in urban communities he first outlined in *Uses of Disorder*, Sennett contends that in 'A city of differences and of fragments of life that do not connect, in such a city the obsessed are set free' (1990: 125). Indeed, he goes on to declare that 'Deviance is the freedom made possible in a crowded city of lightly engaged people' (1990: 126–7). There are similarities here with Iris Marion Young's (1990: 241) arguments for the 'unoppressive city' where there is 'an openness to unassimilated others' and with Jane Jacobs (1961: 83) who, some 30 years before, celebrated 'the tolerance, the room for great differences amongst neighbours' within New York City. For Jacobs, however, it was not indifference that was the key to 'allowing strangers to dwell in peace together on civilised but essentially dignified and reserved terms' but the practices of 'do-it-yourself surveillance' (1961: 121), 'that intricate, almost unconscious network of voluntary controls ... enforced by the people themselves' (1961: 31).

For William Bratton neither Sennett's indifference nor Jacobs 'd-i-y' surveillance are sufficient to establish the city as a tolerant, emancipatory place.

The deviance he detected in his journey through New York City requires a more robust response. Following his appointment in 1994 as Commissioner of the New York Police Department (NYPD), Bratton targeted the beggars, the drunks and the vandals with what became known as zero tolerance policing. Many on the political left have despaired at this strategy as an attempt to erase the diversity to be encountered in public spaces, which is so important to Sennett's view of the emancipatory possibilities of the city. According to Neil Smith (1999), for example, Bratton's zero tolerance approach exemplifies 1990s urban revanchism and is 'a chilling sign of potential urban futures'. But for those on the political right, Bratton's tough 'law and order' approach was about emancipation. 'Freedom', declared Rudolph Giuliani, the New York City Mayor who hired Bratton, 'is about the willingness of every single human being to cede to lawful authority a great deal of discretion about what you do' (quoted in Smith, 1999: 189). From this perspective, zero tolerance policing would help 'reclaim the streets for respectable law-abiding people and help overcome the "culture of fear" … characteristic of late modern urban environments' (Hughes, 1998: 112).

In this chapter I want to explore these tensions and anxieties around the inter-play of deviance, difference, and crime control in the city. For Sharon Zukin (1995: 27), 'the democratization of public space' is entangled with peoples' fears for their physical security. She warns that 'One of the most tangible threats to public culture comes from the politics of everyday fear' (1995: 38). Similarly, Ash Amin and Stephen Graham (1999: 16) note that 'the tensions associated with the juxtaposition of difference, perceived or real (such as the fear of crime or violence, racial intolerance, uncertainty and insecurity) often put into question the very definition and usage of the phrase "urban public space"'. Sennett, too, acknowledges this in his walk around New York City. 'There is withdrawal and fear of exposure', he observes, 'as though all differences are as potentially explosive as those between a drug dealer and an ordinary citizen' (1990: 129).

The roots of such fears and the ways in which they create and reinforce exclusion from social life in public spaces are complex. At one level, people may be excluded because of fears prompted by the experience of crime itself or as a result of sub-criminal acts such as racist, sexist, homophobic or ageist harassment (Pain, 2001). But fear can also reflect a more general sense of anxiety engendered through the confrontation of difference and difficulty and which, like the fear of crime, can lead to precautionary behaviour which restricts peoples' social activities, their employment opportunities and their freedom of expression (Bannister and Fyfe, 2001). However, it is not just fear and crime that constitute the 'new omens of urban calamity' (Baeten, 2001: 4). Often crime control itself is represented now as a major threat to difference and diversity in the city. Criminologists and left-leaning urbanists make much of how innovations in policing and crime prevention, like zero tolerance policing (ZTP) and closed circuit television (CCTV) surveillance, can be seen 'as tightening the ratchet of social control and as the forerunner of some new technologically sophisticated totalitarianism' (Young, 1999: 90). While supporters of such new crime control strategies believe they can bring about dramatic reductions in crime and the fear of crime, critics view them as little more than 'attempts to purify the public sphere of disorder and difference through the spatial exclusion of those social groups who are judged to be deviant, imperfect and marginal in public space' (Toon, 2000: 141).

These competing perceptions of crime control tactics and technologies have very different implications for our understanding of the emancipatory possibilities of the contemporary city. While proponents of ZTP and CCTV surveillance assert that reductions in fear mean people will enjoy a new sense of freedom in the city, opponents counter with claims that these initiatives are part of an oppressive 'criminology of intolerance' in which the prevailing concern is to exclude anyone 'that will disrupt the smooth running of the system' (Young, 1999: 46). In this chapter I explore these differing interpretations of crime control in the late modern city. Using ZTP and CCTV surveillance as case studies, I will argue that the nature and impacts of ZTP and CCTV surveillance appear far more complex than either their supporters or critics will allow for. To put ZTP and CCTV in context, however, it is first necessary to sketch out the emerging contours of crime control in the late modern city.

Visions and realities: crime control in the late modern city

According to David Garland, crime control in the UK and USA, currently exhibits two new and distinct lines of governmental action: 'an *adaptive strategy* stressing prevention and partnership and a *sovereign state strategy* emphasising enhanced control and expressive punishment' (2000: 348). The adaptive strategy involves the state withdrawing its claim to be the chief provider of security and instead emphasizes the use of preventive techniques, like increased surveillance, and the role of public-private partnerships in areas like policing and urban design. Yet, Garland argues, governments are 'deeply ambivalent' about this new infrastructure of crime prevention. At certain times in particular spaces and with respect to specific offences and offenders, government's reactivate 'the old myth of the sovereign state' with a strategy of more intensive modes of policing and punishment, like harsher sentencing, greater use of imprisonment and zero tolerance policing.

The reasons behind these new approaches to crime control are rooted in the complex social, economic, and political changes associated with late modernity. These include fiscal pressures on the state, the emergence of new forms of governance and the rise of new technologies, and, of particular importance here, a 'new collective experience of crime and insecurity' (Garland, 2000: 354). High crime rates have now become a normal social fact. Jock Young (1999: 122) notes how 'crime has moved from the rare, the abnormal, the offence of the marginal and the stranger, to a commonplace part of the texture of everyday life ... as well as extending its anxiety into all areas of the city'. Of course, the collective experience of crime remains highly differentiated both socially and spatially but significantly it is the middle classes who since the 1980s have experienced some of the most dramatic changes in the way crime features in their lives. 'From being a problem that mostly affected the poor, crime (and particularly vandalism, theft, burglary and robbery)' Garland contends, 'increasingly became a daily consideration for anyone who owned a car, used a subway, left their home unguarded during the day, or walked the city streets at night' (2000: 359).

Against this background of changes in the frequency and distribution of crime, the middle classes have 'scripted the official response to crime rates' (Smith, 1999: 199), resulting in new approaches to crime control that have left their mark in cities on both sides of the Atlantic. The adaptive strategy of prevention and partnership is evident in the growth in community policing initiatives, the proliferation of neighbourhood watch schemes in affluent sub-urbs, and the encouragement given to the 'fortress impulse' in architecture and urban design. Susan Christopherson (1994: 421), for example, describes cities dominated by 'security cages and a honeycomb of residential and busi-ness fortress', while Eugene McLaughlin and John Muncie (2000: 117) write of cities now riddled with 'sharply demarcated privatised walled and gated enclaves'. Traditional reliance on the policing by the state has given way to new 'security networks' (Newburn, 2001) based around private forms of policing and the proliferation of electronic surveillance. In terms of the 'sover-eign state strategy', the repertoire of initiatives now being deployed in the US and UK is no less extensive: 'Harsher sentencing and the increased use of imprisonment; ... the revival of chain gangs and corporal punishment; boot camps and "supermax" prisons; ... community notification laws and pae-dophile registers; zero tolerance policing and sex offender orders' (Garland, 2000: 350). Although the visible imprint of this strategy might be less than the adaptive strategy, it is no less important to the fortunes of contemporary cities:

> In the world of global finances, state governments are allotted the role of little else than oversized police precincts; the quantity and quality of the policemen [sic] on the beat, efficiency displayed in sweeping the streets clean of beg-gars, pesterers and pilferers, and the tightness of the jail walls loom large among the factors of 'investors' confidence', and so among the items calcu-lated when the decisions to invest or cut the losses and run are made. (Bauman, 2000: 216)

In the context of this chapter, the key questions raised by these two approaches to crime control concern their implications for regulating differ-ence and deviance in the late modern city. Using ZTP and CCTV surveillance as case studies which exemplify the sovereign state and adaptive approaches, I will examine the contrasting discursive constructions of these tactics and technologies of crime control and their significance for anxieties about crime and anxieties about crime control in the city.

Zero tolerance policing (ZTP)

ZTP first came to public attention in the mid-1990s with the appointment of William Bratton as the Commissioner of the New York City Police Department (NYPD), where he introduced a policing strategy of targeting 'quality of life' offences. These so-called 'beer and piss' patrols focused on drunkenness, public urination, begging, vandalism, and other anti-social behaviour. Entitled 'Reclaiming the Public Spaces of New York' (New York City Police Department, 1994), this strategy was 'the linchpin of efforts ... being undertaken by the New York City Police Department to reduce crime and the fear of crime in the city' (Silverman and Della-Giustina, 2001: 950). The strategy was based on the claim that quality of life offences, like aggressive begging, squeegee cleaners, street prostitution, boombox cars, public drunkenness, reckless bicyclists and graffiti, restricted the use of public space and contributed to 'the sense that the entire public environment is a threatening place' (New York City Police Department, 1994: 5). The rationale behind focusing on quality of life offences was that 'strong and authoritative use of coercive police powers' in relation to these offences would reduce fear and prevent more serious types of disorder and crime from occurring (Innes, 1999, p. 398). This reasoning was leant academic credibility by the 'broken windows' thesis advanced by James Q. Wilson and George L. Kelling (1982):

> The presence of signs of dilapidation can instigate a feedback cycle, in which fear is engendered in the law-abiding members of the local population whose informal mechanisms of surveillance and control are the main guarantors of order in the locality. This fear leads to a general retreat from public interaction amongst the law-abiding groups and thus informal controls decrease and criminal and disorderly activities rapidly increase. (Innes, 1999: 398)

Bratton's crackdown on minor offences immediately seemed to deliver encouraging results. In New York City, the recorded crime rate between 1994 and 1997 dropped by 37 per cent, with a fall in homicide of over 50 per cent. According to Bratton (1997: 29) there was a simple explanation for this: 'Blame it on the police.' Not surprisingly, this 'triumphalist discourse' (Body-Gendrot, 2000: 117) quickly began to capture the attention of politicians and police chiefs in other countries with NYPD becoming the most visited and researched police department in the world. In the summer of 1995, the then UK shadow Home Secretary, Jack Straw, visited New York City and met

Bratton. On his return, Straw promised that a Labour government would introduce ZTP and 'reclaim the streets for the law-abiding citizen' from the 'aggressive begging of winos, addicts and squeegee merchants' (quoted in Bowling, 1999: 532). Such moral authoritarianism has since been put into practice by the adoption of zero tolerance style policing in various parts of the UK, including the Kings Cross area in London, Cleveland in north east England and Strathclyde in the west of Scotland. In all these areas, advocates of ZTP have claimed dramatic reductions in levels of crime and the fear of crime (see Dennis and Mallon, 1997; Orr, 1997). Of course, any attempts to link directly falling crime levels to a specific policing initiative are fraught with problems. As Ben Bowling (1999: 551) concluded in New York City, for example, 'only a circumstantial case has been made for the link between aggressive policing and falling crime'. The fall in homicide in the city in the mid-1990s, he argues, had much more to do with the contraction in the crack cocaine market than any innovations in policing strategy. Moreover, in cities which did not adopt ZTP tactics in the US, crime also declined dramatically in the mid-1990s. In San Diego, for example, the policing approach adopted was very different to that in New York City and involved developing partnerships between the police, public and private sectors yet falls in crime have been comparable to those recorded in New York City (Body-Gendrot, 2000).

Despite these uncertainties about its impact, support for ZTP remains among some police officers and politicians in what increasingly appears as 'a last ditch attempt' by the state to 'make policing and legal regulation succeed as a means of social control' (Hirst, 2000: 281). At the same time, however, criticism of ZTP has grown. Indeed, condemnation of ZTP has yielded a surprising alliance of senior police officers, criminologists, and left-leaning urbanists, alarmed by its implications for public order, difference, and diversity in the city. Some UK police officers view ZTP as a return to the failed military-style policing tactics of the 1970s and 1980s, which contributed to serious public disorder in several cities (Hopkins Burke, 1998). For example, Charles Pollard (1997: 44), Chief Constable of the UK's Thames Valley force, is critical of the 'ruthlessness in dealing with low-level criminality and disorderliness ... and of the single-minded pursuit of short term results'. Some criminologists (e.g. Body-Gendrot, 2000; Greene, 1999) have echoed these concerns, highlighting the way in which any reductions in crime have come at a severe cost to public trust in the police, particularly among minority communities. Others, however, go much further. Some describe ZTP as a 'Robocop version of beat policing [which] could quite easily destroy the "ballet of the street" and "benign disorder" that ... are so crucial to a vital street life' (McLauglin and Muncie, 2000: 130; cf. Merrifield, 2000: 484). Smith offers one of the most sustained attacks on ZTP. According to him, the publication of the NYPD's (1994) *Reclaiming the Public Spaces of New York* signalled 'the advent of a *fin de siècle* American revanchism in the urban landscape' involving 'a visceral identification of the culprits, the enemies who had stolen from the white middle class a city that members of the latter assumed to be their birthright' (1999: 187).

These anxieties about the implications of ZTP for difference and diversity are important not least because they challenge the complacent and self-interested claims of supporters of this policing strategy. Nevertheless, the radical totalitarian reading of ZTP offered by some critics, in which crime control rather than crime is constructed as the major threat to urban life and culture, can be as misleading and exaggerated as the utopian claims of advocates of ZTP. While the latter see it as some kind of 'easy miracle and instant cure' (Young, 1999: 130) to the problems of public space, critics simply invert this rhetoric and view ZTP as an unmitigated disaster delivering instant oppression and crushing any 'street spontaneity and vibrancy' (Merrifield, 2000: 485). But this interpretation places too much weight on a caricature of ZTP when even a casual inspection of the growing research literature reveals that ZTP is a far more complex set of practices than its popular image uncritically recycled by many of its opponents. In particular, while critical perspectives on ZTP focus on the everyday life of those on the street in all its kaleidoscopic detail, the police are simply portrayed as an 'abstract and systematic order which is responsible for oppression' (Thrift, 2000: 235).

Examination of police practices in New York City has revealed that enforcement of quality of life crimes served largely to help with intelligence gathering in relation to more serious offences (Greene, 1999). Moreover, rigorous enforcement was only one element in a wider set of reforms of management, communication and intelligence gathering designed to make the police more efficient, effective, and accountable. It is not surprising therefore that researchers like Eli Silverman and Jo-Ann Della-Giustina have concluded that ZTP has generated a series of myths that bare little relation to reality. In particular, they note that critics too often depict zero tolerance as 'leaving the police little or no discretion' when in reality zero tolerance cannot mean '24 hours, 7 days a week, perpetual enforcement of all quality of life offences' (2001: 954). This is even echoed by Bratton who has emphasized that discretion is a vital part of policing and that what happened in New York City was not zero tolerance policing. Indeed, Bratton has roundly condemned the concept, emphasizing that policing involves developing plans in consultation with local communities and taking action against a range of crime and incivilities (Young, 1999: 124). Moreover, critics leave 'little scope for the specifics of locality and difference' (Hughes, 1997: 158).

Comparative research within New York City has revealed, for example, the importance of local variations in zero tolerance policing styles. A study of two precincts in the South Bronx highlighted significant reductions in crime and civilian complaints against the police as a result of local styles of management adopted by police commanders which emphasized improved supervision of officers and greater community interaction (Davis and Mateu-Gelabert, 1999). Moving from the intra-urban to the international scale, there are also clear differences between ZTP in the US and the UK. Despite senior UK police officers engaging with the popular discourse of ZTP, policing in areas that claim to have experimented with ZTP has been orientated towards the cultivation of informants and information-based police action rather than aggressive enforcement. Indeed Johnston (2000: 67) concludes that ZTP in the UK context has more to do with risk-based policing (in the sense of information gathering and proactive intervention) than with addressing the problem of incivility.

There is a second difficulty for those who only see danger in 'the zero tolerance gospel' (Shapiro, 1997). An important insight provided by the broken-windows thesis underpinning zero tolerance strategies is that 'the cumulative effect of minor incivilities poses as much of a problem to the public as crime itself' (Young, 1999: 138). As Young argues, this was recognized over ten years ago when an Edinburgh women's domestic violence campaign first coined the term 'zero tolerance' in relation to violence against women. This demanded explicit recognition that rape or other forms of sexual violence represent 'the end-point of the continuum of aggressive sexual behaviour' (1999: 138) that might also including staring, touching, and verbal sexual abuse. Similar arguments could, of course, be made in relation to the abuses suffered by a range of other vulnerable social groups, including gays and lesbians, the young and the elderly, and ethnic and racial minorities. Few would dispute that failure to adopt a zero tolerance approach in these contexts contributes greatly to the exclusion of individuals and social groups. Some critics acknowledge that a democracy cannot 'allow all types of disorder to run amok', but this poses a dilemma for those who believe they have to 'fend off and contest Zero Tolerance tactics' (Merrifield, 2000: 885): how to differentiate between 'good' and 'bad' zero tolerance. What this suggests is the need for a greater engagement with the ambivalence of ZTP. On the one hand, some forms of zero tolerance may severely inhibit freedoms in the city; as Silverman and Della-Giustina (2001: 954) note, 'When zero tolerance changes into zealous pursuit of all quality of life offences, everyone loses – public and police alike in their ability to fight crime'. On the other hand, zero tolerance of particular incivilities may actually enhance the confidence of some groups in using the city. Thus, while critics of ZTP may wonder 'whether the baby of disorder might be getting ditched with the criminal bathwater' (Merrifield, 2000: 484) it is also important that these same critics don't simply throw the 'smart policing baby' out with 'the aggressive policing' bathwater (Bowling, 1999: 550).

Closed circuit television (CCTV) surveillance

While ZTP is largely associated with the cosmopolitan world of New York City, it is the relatively quiet, seaside town of Bournemouth on England's south coast which regularly features in discussions of CCTV surveillance.

Bournemouth was the birthplace of the UK's 'surveillance revolution' (Williams et al., 2000: 169). Here in August 1985 the country's first public space CCTV system went 'live' and over the next ten years there was the piece-meal establishment of similar schemes in other towns and cities. From the mid-1990s, however, the diffusion of CCTV schemes accelerated. In 1994, 78 towns and cities in the UK had CCTV; by May 1999, 530 town and city centre schemes were either in operation or had funding allocated (Williams et al., 2000: 170). UK cities now have the most intensive concentration of electronic 'eyes on the street' in the world (McLaughlin and Muncie, 2000: 130).

The reasons for this rapid spread of public space CCTV systems are now well documented (Fyfe and Bannister, 1998, Williams et al., 2000). Economically, the spread of CCTV is bound up with urban regeneration agendas and attempts to revive the fortunes of the entrepreneurial city by 'managing out inappropriate behaviour in the new territories of consumption' (McCahill, 1998: 42). As Ian Toon (2000: 150) observes, CCTV is part of 'an attempt to reinvent public spaces, promoting the commercial functions of the street above all other uses'. Politically, the involvement of businesses and local government in financing surveillance schemes demonstrates how CCTV fits an agenda of developing public-private partnerships that enable the central state to pursue its 'responsibilization' strategy of off-loading certain crime control activities onto the local state and the private sector.

As with ZTP, there are no shortage of CCTV supporters willing to claim that this technology has brought considerable benefits to the city. 'Before' and 'After' studies of the impact of CCTV on the incidence of crime in several town and city centres suggest that crime rates have fallen dramatically and contributed to a so-called 'feel good factor' by 'reassuring town centre users that they are safer' (Home Office, 1994: 14). Opinion surveys also routinely reveal high levels of public support: 85 per cent of those asked in Sutton in south-east London said they would be in favour of town centre CCTV, rising to 95 per cent of those asked in Glasgow. CCTV surveillance therefore appears to be a cause for celebration: it 'increases public freedom, enhancing opportunities for people to enjoy public places' (Arlidge, 1994: 22). Yet, the utopian, deterministic rhetoric surrounding CCTV with its promise of a 'feel good factor' for those who use town and city centres is increasingly being questioned. Evidence for reductions in crime and the fear of crime is far from consistent (Fyfe and Bannister, 1998) and concerns remain that CCTV systems simply displace crime to areas out of the view of cameras, areas which may be less able to cope with the problems of crime. Nor is public support for CCTV as high or robust as is often claimed given serious methodological weaknesses in the ways in which opinion surveys have been conducted (Ditton, 2000). More generally, as Roy Coleman and Joe Sim (1998) cogently argue, the 'feel good factor' focuses on a particular conception of danger and disorder in city centre public spaces, marginalizing other areas of crime and anxiety, such as domestic violence experienced in private spaces.

While these uncertainties about the impact of CCTV on crime and deviance appear to have done little to stem popular and political enthusiasm for public space CCTV surveillance, a growing critical discourse of this 'silver bullet' of crime prevention has also developed informed by a similar radical totalitarianism to that underpinning criticism of ZTP. In *The Maximum Surveillance Society*, Clive Norris and Gary Armstrong (1999) note, for example, how the academic literature on CCTV is replete with allusions to Orwell's *Nineteen Eighty Four*, Big Brother, and Foucault's discussion of Bentham's Panopticon. 'The dominant cultural theme is tragic', they conclude, 'in which CCTV is placed in the context of those forms of surveillance which are deeply implicated in the structure of totalitarian rule' (1999: 5). More specifically, many urbanists focus on the contribution of CCTV surveillance to processes of social exclusion and the end of public space. Steven Flusty (1998: 59), writing in the context of Los Angeles, suggests that video cameras are now part of 'an infrastructure restructuring the city into electronically linked islands of privilege embedded in a police state matrix'. Jon Bannister and I have contributed to this apocalyptic vision of CCTV, warning that under the constant gaze of CCTV surveillance cameras, any claim that streets symbolize public life with all its human contact, conflict and tolerance would be difficult to sustain (Fyfe and Bannister, 1998: 265). More recently, Williams et al. (2000: 184) declared that CCTV is being used as an instrument to preserve 'the public spaces of our town centres ... for the consumer citizen, while those whose spending power is low ... are effectively excluded'.

Yet, as with ZTP, this dystopian vision of CCTV risks offering a misleading and exaggerated account of its implications for difference and diversity in the late modern city. The temptation to draw upon Foucault's reading of panopticism as a way of understanding the impacts of CCTV, for example, needs to be treated with caution. While there are some similarities, city streets are not the same as the bounded institutional spaces of the prison or asylum. Unlike prisoners and asylum inmates, people in cities do not 'suffer continuous confinement' (Hannah, 1997: 344), nor are those using the streets always aware they are being monitored with CCTV, whereas in the Panopticon the certainty of detection and therefore intervention meant conformity was a necessity. In CCTV systems detection and intervention are contingent on a range of social processes, including whether the screens are being monitored and whether an activity defined as deviant actually generates a response (Norris and Armstrong, 1999: 92). Indeed, opponents of CCTV, just as much as supporters, appear to share an 'unthinking belief that, such is the impressive technological power of those cameras ... that little or no human effort has to be added for them to be highly efficacious' (Ditton and Short, 1999: 39). This position is untenable. CCTV, like other technological innovations, doesn't create 'an electronic world swept of people' but 'hybrid "actor-networks" of people and electronic things' (Thrift, 1996: 1473).

Research into these 'actor-networks' of CCTV cameras, control room operators, and police officers is relatively rare. While critics are keen to revel in the exclusionary experiences of those subject to the gaze of the cameras, they are far less interested in understanding 'how the information generated by CCTV systems is selected, evaluated, used, and acted upon' (McCahill, 1998: 46). One study of CCTV surveillance frankly admits that 'the intention ... was not to collect the accounts of police officers and CCTV monitoring staff themselves on how they police teenagers in public environments, but rather to provide an account of prohibitions and constraints on activities from the point of view of the excluded' (Toon, 2000: 163). The latter is, of course, important, but so too (if one is to avoid the trap of technological determinacy) is the role of those operating CCTV systems. In one of the few published studies of the activities and inter-relationships between CCTV control room staff and police officers, Norris and Armstrong found that the deployment of police officers in response to activities witnessed in CCTV control rooms is a rare event. In 600 hours of observations in three control rooms there were only 45 deployments most of which were in response to crimes of violence. As these researchers conclude, 'Authoritative intervention is a relatively rare phenomenon, and few incidents result in deployment, fewer still in arrest' (1999: 168). The reasons for this are complex and reflect the interplay of organisational, structural and personal factors. CCTV operators can only request, rather than demand, police intervention and must be able to provide a robust justification for requesting police action. Indeed, without a specific legal mandate to intervene, the police are reluctant to become involved. For Norris and Armstrong (1999: 200) this suggests that the 'exclusionary potential' of CCTV, which has so concerned critics, is not particularly marked in practice, 'primarily because deployment [is] such a rare feature'.

The fact that intervention and arrest are relatively rare does not, of course, mean that 'significant social interaction, albeit remote and technologically mediated, has not taken place' (Norris and Armstrong, 1999: 151). For example, a study of young peoples' responses to the introduction of town centre CCTV in England revealed how they felt forced to find 'concealed interstitial spaces within, and "invisible" routeways through, the town centre' in order to escape the gaze of CCTV and 'reappropriate space for themselves' (Toon, 2000: 154). Nevertheless, it is also clear that CCTV surveillance, like ZTP, is 'much messier, contingent and open to contested interpretations than is implied in much of the literature to date' (Hughes et al., 2001: 333). Thus, rather than portraying CCTV surveillance in either utopian terms (as the silver bullet of crime control) or as part of a starkly dystopian discourse (of an Orwellian future of totalitarian control), a more nuanced position is required. Indeed, ambiguity and ambivalence are precisely the characteristics that now appear to inform local peoples' responses to the proliferation of CCTV. There is 'a grudging acceptance of the "need" for video surveillance with a range of (often diffuse) worries about how CCTV might come to be used and about the kind of world it signifies' (Sparks et al., 2001: 894). A sense of ambivalence is also central to Hille Koskela's examination of CCTV surveillance as an emotional experience:

> To be under surveillance is an ambivalent emotional event. A surveillance camera, as an object, can at the same time represent safety and danger. To be protected can feel the same as being threatened. A paradox of emotional space is that it does, indeed, make sense that surveillance cameras can make people feel both more secure and more fearful. (2000: 259)

In more concrete terms Young (1999: 193) makes the point that while CCTV is 'one of the most invidious of inventions' and something which can make Orwell's *Nineteen Eighty Four* a reality, in a different political context, it can be liberating and protective: 'The cameras can be turned around; their context and control can be changed'. Examples of the positive uses of CCTV include Newham in east London where it has been employed to provide protection against racist attacks using facial recognition software in order to identify known racist offenders on the street (Lyon, 2001: 58–9). Another is the way CCTV changes the ability of police officers to privilege their accounts of events given that cameras can now monitor their performance and activity on the street (Norris and Armstrong, 1999: 188). Rather than simply condemning CCTV as 'a product of some capitalist conspiracy or the evil effects of a plutocratic urge' (Lyon, 2001: 2) or uncritically accepting it as a 'silver bullet' of crime prevention, the real challenge is to find ways in which 'the new technologies of mass surveillance can be harnessed to encourage participation rather than exclusion, strengthen personhood rather than diminish it, and be used for benevolent rather than malign purposes' (Norris and Armstrong, 1999: 230).

Conclusions: futures of crime control

Debates over the tactics and technologies of crime control are increasingly polarized between utopian and dystopian visions of their implications for difference and deviance in the late modern city. Intriguingly, these debates have distinct echoes of arguments among historians about crime control in the modern city of the nineteenth century. 'Orthodox' accounts of the introduction of formal policing into urban areas as a means of 'solving the problem of order and checking the spread of lawlessness' (Reiner, 1992: 19) are pitched against 'revisionist' claims that the police were part of a 'centralised social system in which the state penetrated the depths of society' (1992: 34). Just as these revisionist accounts constitute an advance in understanding beyond orthodox views of the modern police, so too those critical of late modern methods of crime control have significantly expanded our understanding of ZTP and CCTV. Nevertheless, critics have often failed to engage with the complex and open-ended nature of crime control in the late modern city. Indeed, the radical totalitarianism which informs many of these critical perspectives tends to 'exaggerate the dystopian tendencies at work and the power of the intrusive social control machine' (Hughes, 1997: 157). As Gordon Hughes wryly notes, 'Glimpses of dystopia have a powerful appeal, not least to intellectuals who, doubtless, gain vicarious pleasure from being on the "edge", compared to the supposedly slumbering masses' (1997: 158). Yet the problem remains of how to address the tensions between anxieties about crime and anxieties about crime control at a time when crime is 'no longer a marginal concern, an unexpected incident in their life, but an ever-present possibility' (Young, 1999: 36). In part this means engaging with the ambivalence of crime control and the paradoxes and possibilities it entails.

It is also important to think beyond current approaches to crime control and look for ways outside the criminal justice system of addressing questions of deviance and diversity in the city. In this regard there is a growing European radical communitarian vision of tackling crime by rethinking political and economic democracy to develop a progressive agenda of social inclusion and pluralism. Of particular importance here is the work of Paul Hirst (2000). According to Hirst, current strategies of crime control attempt to enforce hierarchical control on an increasingly heterogeneous and unequal society and to develop systems of surveillance based on norms that are increasingly contested. This results in a system of criminal justice that is both ineffective and inappropriate. In its place, Hirst argues for an approach to crime control informed by associationalism, of a society split into self-governing communities with different values and with publicly funded services devolved to self-governing voluntary associations. Groups would co-exist yet keep their own values by a mixture of micro-governance (special zones where different rules apply) and mutual extra-territoriality (self-governing communities sharing the same space but applying rules in matters of community concern to their members alone). As Hirst speculates:

> Imagine cities clearly divided into permissive and restrictive zones with regard to drug use. Imagine that if the rich can live in gated communities with security guards, that the metaphorical 'ghettos' of the USA became more like real ones with their own boundaries and their own local policing. Tell that to the LAPD? But the cost of their rule is immense, including the cost of riots, and its results ineffective. (2000: 291)

Elements of such strategies already exist in some cities. The Danes tolerate the anarchist enclave of Christiania in Copenhagen, and in the Netherlands there are marginal zones where prostitution and soft drug use are tolerated. Rather than looking for approaches to crime control that apply across the city and are delivered by one central agency, there needs to be a commitment to developing more diverse, decentralised and self-regulatory strategies. As Hirst explains:

Outside a thin core of public morality – almost everyone will agree that murder, theft and fraud are crimes – groups will be better off setting and policing their own standards. Citizens would have to accept that different rules applied to different communities with informal self-regulation and arbitration. (2000: 279)

Intriguingly this view takes us back to the streets of New York City with which this chapter began. According to Jane Jacobs (1961: 31), probably the City's most infamous flaneur, the 'public peace' of cities is not kept primarily by the police but by 'an intricate, almost unconscious, network of voluntary controls and standards among the people themselves and enforced by the people themselves'. Although this might be dismissed as a utopian vision of crime control in the late modern city, is it any less utopian than the belief that we can control crime by 'changing the way it is handled within the formal system of criminal justice' (Currie, 1985: 229)?

References

Amin, A. and Graham, S. (1999) 'Cities of connection and disconnection', in J. Allen, D. Massey and M. Pryke (eds), *Unsettling Cities*. London: Routledge. pp. 7–47.

Arlidge, J. (1994) 'Welcome big brother', *The Independent*, 2 November, p. 22.

Baeten, G. (2001) 'Hypochondriac geographies of the city and the new urban dystopia: coming to terms with the "other" city'. Paper presented at the American Association of Geographers Conference, New York.

Bannister, J. and Fyfe, N. R. (2001) 'Fear and the city', *Urban Studies*, 38: 807–14.

Bauman, Z. (2000) 'Social issues of law and order', *British Journal of Criminology*, 40: 205–21.

Body-Gendrot, S. (2000) *The Social Control of Cities? A Comparative Perspective*. Oxford: Blackwell.

Bowling, B. (1999) 'The rise and fall of New York murder: zero tolerance or crack's decline?', *British Journal of Criminology*, 39: 531–54.

Bratton, W.J. (1997) 'Crime is down in New York City: blame the police', in N. Dennis (ed.), *Zero Tolerance: Policing a Free Society*. London: Institute of Economic Affairs. pp. 29–42.

Christopherson, S. (1994) 'Fortress city: privatised spaces, consumer citizenship', in A. Amin (ed.), *Post-Fordism: A Reader*. Oxford: Basil Blackwell. pp. 409–27.

Coleman, R. and Sim, J. (1998) 'From the docklands to the Disney store: surveillance, risk and security in Liverpool City Centre', *International Review of Law, Computers and Technology*, 12: 45–63.

Currie, E. (1985) *Confronting Crime: An American Challenge*. New York, Pantheon Books.

Davis, R.C. and Mateu-Gelabert, P. (1999) *Respectful and Effective Policing: Two Examples in the South Bronx*. New York: Vera Institute.

Dennis, N. and Mallon, R. (1997) 'Confident policing in Hartlepool', in N. Dennis (ed.), *Zero Tolerance: Policing a Free Society*, London: Institute of Economic Affairs. pp. 61–86.

Ditton, J. (2000) 'Public attitudes towards open-street CCTV in Glasgow', *British Journal of Criminology*, 40: 692–709.

Ditton, J. and Short, E. (1998) 'Evaluating Scotland's first town centres CCTV scheme', in C. Norris, J. Moran and G. Armstrong (eds), *Surveillance, CCTV and Social Control*. Aldershot: Ashgate. pp. 155-74.

Ditton, J. and Short, E. (1999) *The Effect of CCTV on Recorded Crime Rates and Public Concern about Crime in Glasgow*. Edinburgh: Scottish Office.

Flusty, S. (1998) 'Building paranoia', in N. Ellin (ed.), *Architecture of Fear*. Princeton, NJ: Princeton University Press. pp. 47–59.

Fyfe, N.R. and Bannister, J. (1998) ' "The eyes upon the street": closed circuit television surveillance and the city', in N. R. Fyfe (ed.), *Images of the Street: Planning, Identity and Control in Public Space*. London: Routledge. pp. 254–67.

Garland, D. (2000) 'The culture of high crime societies: some preconditions of recent "law and order" policies', *British Journal of Criminology* 40: 347–75.

Greene, J. (1999) 'Zero tolerance: a case study of police policies and practices in New York City', *Crime and Delinquency*, 45: 172–87.

Hannah, M. (1997) 'Imperfect panopticism: envisioning the construction of normal lives', in G. Benko and E. Strohmayer (Eds.), *Space and Social Theory: Interpreting Modernity and Postmodernity*. Oxford: Blackwell. pp. 344–59.

Hirst, P. (2000) 'Statism, pluralism and social control', *British Journal of Criminology*, 20: 279–95.

Home Office (1994) *CCTV: Looking Out For You*. London: Home Office.

Hopkins Burke, R. (1998) 'A contextualisation of zero tolerance policing strategies', in R. Hopkins Burke (ed.), *Zero Tolerance Policing*. London: Perpetuity Press. pp. 11–38.

Hughes, G. (1997) 'Policing late modernity: crime management in contemporary Britain', in N. Jewson and S. Macgregor (eds), *Transforming Cities: Contested Governance and New Spatial Divisions*. London: Routledge. pp. 153–65.

Hughes, G. (1998) *Understanding Crime Prevention: Social Control, Risk, and Late Modernity*. Buckingham: Open University Press.

Hughes, G., McLaughlin, E. and Muncie, J. (2001) 'Teetering on the edge: futures of crime control and community safety', in G. Hughes, E. McLaughlin and J. Muncie (eds), *Crime Prevention and Community Safety: New Directions*. London: Sage. pp. 318–40.

Innes, M. (1999) ' "An iron fist in an iron glove?" the zero tolerance policing debate', *The Howard Journal*, 38: 397–410.

Jacobs, J. (1961) *The Death and Life of Great American Cities: The Failure of Town Planning*. Harmandsworth: Penguin.

Johnston, L. (2000) *Policing Britain: Risk, Security and Governance*. Harlow: Longman.

Koskela, H. (2000) ' "The gaze without eyes": video-surveillance and the changing nature of urban space', *Progress in Human Geography*, 24: 243–65.

Lyon, D. (2001) *Surveillance Society: Monitoring Everyday Life*. Buckingham: Open University Press.

McCahill, M. (1998) 'Beyond Foucault: towards a contemporary theory of surveillance', in C. Norris, J. Moran and G. Armstrong (eds), *Surveillance, Closed Circuit Television and Social Control*. Aldershot: Ashgate. pp. 41–65.

McLaughlin, E. and Muncie, J. (2000) 'Walled cities: surveillance, regulation and segregation', in S. Pile, C. Brook, C. and G. Mooney (eds), *Unruly Cities?* London: Routledge. pp. 103–48.

Merrifield, A. (2000) 'The dialectics of dystopia: disorder and zero tolerance in the city', *International Journal of Urban and Regional Research*, 24: 473–88.

Newburn, T. (2001) 'The commodification of policing: security networks in the late modern city', *Urban Studies*, 38: 829–48.

New York City Police Department (NYPD) (1994) *Strategy Number Five: Reclaiming the Public Spaces of New York*. New York: City of New York.

Norris, C. and Armstrong, G. (1999) *The Maximum Surveillance Society: The Rise of CCTV*. Oxford: Berg.

Orr, J. (1997) 'Strathclyde's spotlight initiative', in N. Dennis (ed.), *Zero Tolerance*. London: Institute of Economic Affairs. pp. 104–23.

Pain, R. (2001) 'Gender, race, age and fear in the city', *Urban Studies* 38: 899–913.

Pollard, C. (1997) 'Zero tolerance: short term fix, long term liability?', in N. Dennis (ed.), *Zero Tolerance*. London: Institute of Economic Affairs, pp. 43–57.

Reiner, R. (1992) *The Politics of the Police*. London: Harvester.

Sennett, R. (1970) *The Uses of Disorder: Personal Identity and City Life*. London: Faber and Faber.

Sennett, R. (1990) *The Conscience of the Eye: The Design and Social Life of Cities*. London: Faber and Faber.

Shapiro, B. (1997) 'Zero tolerance gospel', *Index on Censorship*, 4: 17–23.

Silverman, E. and Della-Giustina, J.-A. (2001) 'Urban policing and the fear of crime', *Urban Studies*, 38: 941–58.

Smith, N. (1999) 'Which new urbanism? New York City and the revanchist 1990s', in R.A. Beauregard and S. Body-Gendrot (eds), *The Urban Moment: Cosmopolitan Essays on the Late 20th-century City*. Thousand Oaks, CA: Sage. pp. 185–208.

Sparks, R., Girling, E. and Loader, I. (2001) 'Fear and everyday urban lives', *Urban Studies*, 38: 885–98.

Thrift, N. (1996) 'Old technological fears and new urban eras: reconfiguring the good-will of electronic things', *Urban Studies*, 33: 1463–93.

Thrift, N. (2000) '"Not a straight line but a curve", or, cities are not mirrors of modernity', in D. Bell and A. Haddour (eds), *City Visions*. Harlow: Longman. pp. 233–63.

Toon, I. (2000) 'Finding a place in the street': CCTV surveillance and young people's use of public space', in D. Bell and A. Haddour (eds), *City Visions*. Harlow: Longman. pp. 141–65.

Williams, K. S., Johnstone, C. and Goodwin, M. (2000) 'CCTV surveillance in urban Britain: beyond the rhetoric of crime prevention', in J. R. Gold and G. Revill (eds), *Landscapes of Defence*. London: Prentice Hall. pp. 168–87.

Wilson, J. Q. and Kelling, G. L. (1982) 'Broken windows', *Atlantic Monthly*, March: pp. 29–38.

Young, I. M. (1990) *Justice and the Politics of Difference*. Princeton, NJ: Princeton University Press.

Young, J. (1999) *The Exclusive Society: Social Exclusion, Crime and Difference in Late Modernity*. London: Sage.

Zukin, S. (1995) *The Cultures of Cities*. Oxford: Basil Blackwell.

4 Impurity and the Emancipatory City: young people, community safety and racial danger

Les Back and Michael Keith

The city is normally considered emancipatory because of the potential of its physical, political, and social complexity to accommodate difference, facilitate intercultural dialogue, and promote citizenship. Consequently, the notion of an emancipatory city depends on the realization of particular forms of progressive spatiality. But in the often melancholic subtext of urban social theory, a nostalgia for the permissive public spaces of old persists alongside critiques of the putatively inauthentic or repressive public spaces of the contemporary shopping mall, the themed urban experience, or the dystopian streets of our surveillance society.

In this chapter we are interested in thinking about how the nature of government and of urban space is changing in the context of various moral concerns over racial tension, community safety, and the behaviour of young people. And in that context the trouble with the melancholic narratives of urban social theory about the world we have lost is that they fail to address two major forces transforming the cities of everyday life. First, both the city and the communities inhabiting it are now becoming the objects of new governmental regimes designed to regulate the conduct of conduct through localized and participatory mappings of urban space and the objects within it on planes of risk. Second, the institution of these new regimes of governmentality is locally mediated by story telling and other forms of behaviour that reinvent the nature of everyday places in the city and transform the parameters of acceptable conduct. These popular reimaginings are sometimes progressive, but at other times reiterate racialized conflict and violence thus confounding the populist instinct in urban social theory to lionize (progressive) civil society against the (repressive) state.

Drawing on interviews and other material collected as part of a study of social policy initiatives in the Isle of Dogs and Deptford in inner London, we want to advance five related arguments.[1] First, we suggest that a new policy agenda in the overdeveloped 'North' – loosely associated with Third Way politics on both sides of the Atlantic – demands a specific *localization* of governmental practices that implicitly reinvents a particularly communitarian notion of public space. Second, we suggest that the localities of the city might be considered in terms of the specific *spaces of governmentality* through which the myriad of policy professionals and social policy organisations focus on and attempt to affect particular forms of social behaviour. Our third proposition is that it is helpful to think about the landscapes of the city in terms of the *micro-public spheres* of specific buildings, sites, and places associated with routinized forms of behaviour structuring the temporality of social processes. At times in a mundane fashion, and at others more profoundly, they symbolize the lived, remembered, and forgotten *histories* of particular places. They are also often the subjects of 'institutional rumour and gossip' that invoke variations on these local histories as a narrative frame to render the geographical present comprehensible. Fourth, we argue that the stories woven around specific sites and places in the Isle of Dogs and Deptford exemplify broader processes by which the local disrupts the localizing practices at the heart of the new regimes of government. Finally, our specific discussion of these two areas on either side of the River Thames in London highlights for us the importance of thinking carefully about the sorts of times and places that both provoke racial antagonism and promote intercultural dialogue.

The localizing spaces and subjects of governmentality

Taking as its starting points the imperative to regulate capitalism and the failure of state socialism, Anthony Giddens's (1998) influential *The Third Way* provides an influential diagnosis of the liberal democratic project closely associated with the governments of Bill Clinton and Tony Blair (Mulgan, 1994). Giddens suggested that across the world of late capitalism we are witnessing the emergence of a new democratic state without enemies. Though looking increasingly implausible by 2003, Giddens's (1998: 77) vision for a state without enemies promises to renew the public sphere and produce a double democratization (of both state and civil society institutions) through the mechanisms of direct democracy and devolution coupled with greater transparency, administrative efficiency, and a new role for the state as manager of risk rather than provider of security. As Giddens explains:

> An open public sphere is as important at local as at national level, and is one way in which democratization connects directly with community development. Without it, schemes of community renewal risk separating the community from the wider society. 'Public' here includes physical public space. The degeneration of local communities is usually marked not only by general dilapidation, but by the disappearance of safe public space – streets, squares, parks and other areas where people can feel secure. (1998: 85)

Giddens's (1998: 78) recipe for democratic renewal appeals, in particular, to the notion of 'self-government'. The operative term here is the *self* in self-government. What kind of *self* is to govern itself as the state reinvents the public sphere as an arena of self-regulation? This language of self-government demands careful scrutiny. Self-regulating forms of subjectivity have a particular history that cannot be reduced to a simple movement of something called 'power' away from the state and towards 'the people'. Many critics identify their emergence across the globe with new forms of economic government. In response to the perils of globalization, the state is divesting itself of the responsibility for 'progress' through an increasing stress on the responsibility of self-governing economic actors, whose actions must take as a first principle the rudiments of economic competence (Gordon, 1991; Rose, 1989; 1999).

One major context in which these new, self-regulating approaches to governance are crystalizing is in the neo-liberal state's approach to 'community safety'. It seeks to regulate 'anti-social behaviour' through curfews, individualized legal injunctions (such as anti-social behaviour orders in the United Kingdom), and the criminalization of marginal forms of social interaction, such as begging and sleeping or drinking in public. Critics have seized on these reforms as hallmarks of a new 'revanchist city' (Mitchell, 2003; Smith, 1996). But both conventional and politically radical urban social theorists have paid less attention to the manner in which such approaches to crime reconfigure the relationship between state and civil society through the localization of governmental scrutiny of the 'dangerous' parts of the social. In the rhetoric of self-government exemplified by Giddens (1998), cities are imagined as landscapes of risk. Within these imagined worlds the term community, sometimes with attendant communitarian imperatives, mediates the relationship between state and civil society.

In the United Kingdom, the ongoing transformation of 'public service' delivery institutionalizes this governmental focus on self-regulating forms of community (e.g. DoETR, 1997). For instance, local authorities are now required to work in partnership with other agencies, including the police, to develop crime reduction strategies, youth justice audits, Youth Offender Teams, and Youth Justice Boards. Through these collaborative processes, each local authority is expected to inscribe the spaces of their jurisdiction in terms of a cartography of risk that marks out urban hot spots, risky spaces, and places of crime and disorder (Johnston, 2000). Within this officially sanctioned geography of social policy, young people are often taken to personify the risks of the city, and local authorities are given powers to issue Child Safety Orders and Parenting Orders to restrain their activity.

The project of community safety and crime reduction through self-government contains within it both a moralizing reinvention of individual selves and also some straightforward disciplinary measures, such as youth curfews, to create time-spaces in the public sphere where young people will not be allowed. This criminalization of various forms and norms of behaviour in the United Kingdom both speaks to and draws on similar experiences in the American city in the past decade (e.g. Lees, 2003). But something more than criminality is at stake here. We are witnessing some major changes in the institutions involved in the socialization of young people. In this context the simple division between what is and what is not the state is perhaps not very helpful when the legislative changes relate much more to a transactional relationship between forms and norms of behaviour and of official sanction. Official sanction is now seeking to produce well-behaved young people as much through the activities of particular arms of welfare support as through the punitive institutions of criminal justice. Thus current social policy reforms are as much about what it means to be a young person and what are the rights and responsibilities of such *subjects* as they are about the straightforward relationship between authority and community.

These reforms of government reinvent a notion of public space as a site of civility and possibly even of civilization. Arguably, they are localizing a particular communitarian project (Driver and Martell, 1997). Such a vision of localized self-government rests on an appeal to community that sits uneasily with the cultural complexities of most British inner cities and certainly with the sites of our research. What legislative changes mean for the rights of people who are coming of age or for people whose access to public space has been shaped by racism, poverty, and fear, may be moot in its detail but is certain to be profound in its substance. According to most analyses, intercultural dialogue is deeply dependent on the stages on which it takes place. If the public spaces of the city have conventionally been the sites of what Les Back (1995) has described as the metropolitan paradox in which they are simultaneously sites of inter-cultural dialogue and racist intolerance, then the freedom of the city in a multi-racial context is currently being redefined through the reinvention of its public spaces.

Maps and narratives of the micropublic sphere

The Isle of Dogs and Deptford where we carried out our research are old riverside areas of London on the north and south sides of the River Thames. Both have seen successive waves of neighbourhood change and transnational migration as well as a succession of policy instruments and initiatives aimed at reducing inner city poverty and deindustrialization. In turn these initiatives have created a specific form of institutional landscape in the name of 'urban regeneration' (Keith, 1995).

The argument we would make here is that this institutional landscape shapes the definition of inner city subjects. Funding agencies legitimize the existence of some groups and not others; some forms of social provision receive revenue support and others disappear. Accordingly the boundaries between what is and what is not 'the state' are being renegotiated as funding agencies make visible what is and what is not sanctioned and funded within what is conventionally understood as 'civil society'. Specifically, in the late 1990s the activity of welfare professionals responded to 'the risk agenda' prioritizing funding around 'community safety' and social exclusion'. In the words of one youth worker:

> the rush for everyone to describe everything they're doing in terms of social exclusion and the combating of it ... [is] the same ... as a couple of years ago, [when] everyone suddenly had spent their whole life insuring the safety of the community. And before that everything you did was about skilling and training and access to employment ... The professionals ... refilter the debates on the ground through whatever lens we're using at any given moment.

Such comments point to an alternative history of community safety initiatives in which the national interest in 'community safety' as an object of governmental concern is translated into local initiatives and reshaped at the local level. It is not the principal function of this chapter to explore such a history. But it is important to remember that when 'policy elites' are imagining landscapes of safety and danger they are influenced in part by the regimes of inner city government, which are the institutional realization of the boundary between city government and a remoulded civil society.

The formal territorialization of safety through regimes of localizing governmentality and regulation of conduct translates into both officially sanctioned maps of localizing state intervention and informal narratives of racial fear and danger. Particular sites become emblematic of more substantive issues. On the Isle of Dogs certain estates were seen locally as 'Bengali estates',[2] characteristic of a racialized distinction between Bengali and non-Bengali housing that itself reveals much about ways of thinking about and representing social housing provision. By contrast in Deptford particular places were referred to as 'beyond the state' and the rumoured site of particular forms of criminal activity. The 'ghetto' in the young people's vernacular in South London was used to describe graphically a particular set of estates. In both areas the racialization of space did not correspond exactly to the demographics of settlement but instead conflated reputation and residential presence. We discuss these narrativizations of space by local youth in more detail below.

The point we want to make here is that the professionals of the localizing states are no less subject to the informal storytelling practice of what we're calling the micro-public sphere. The way some places feature in the narratives of policy elites suggests that their formal location on the officially sanctioned cartographies of risk and danger is influenced, in part, by their reputation within the more informal circuits of rumour and gossip within the respective public sector organisations. In interviews informants would not only use a particular estate to exemplify either racial antagonism or problems of racial harassment, they would also refer to the reputation (past and present) of such an estate for particular forms of behaviour. Some youth clubs are identified as the sites of criminal activity, gang behaviour, or drug dealing. Others have a reputation of cross-cultural working. Particular events are said to have precipitated changes in the thinking of the local authority, the police service or community activists. The dictates of space prevent a full analysis, but it might be possible to imagine how a gradual accretion of these stories would produce an iconography of the micro-scale public spheres of the city, a spatial semiotics of the everyday knowledges of welfare professionals.

Significantly, we are not suggesting anything about the empirical accuracy of these accounts of community safety and danger in our two case study areas. Rather we want to suggest such informal anecdotes about particular clubs, streets, and occasions serve, in part, as nuclei around which official cartographies of much wider areas coalesce. At this micro-scale we would suggest that policy actions are shaped by this narrative web. Such storytelling literally constructs the local through processes of narration. It is in this sense that we want to suggest that there is a 'micro-public sphere' of debate that links actions to reputations around the specific times and spaces of the city.

In this way what we are working towards is a rethinking of how *the local* is constituted. Such narrative constructions of the local do not sit easily with the sorts of localizing practices of governmentality emerging from the contemporary political agenda of social reform. Such narrative constructions render space and place as plural, opaque, contested, and frequently contradictory. In contrast the localizing practices of governmental imperatives render the spaces of the city as translucent, uniform, communicatively consistent, and ethically purified. Nor do they accord with Richard Sennett's (2000: 382) more recent typology of reflections on the public realm, in which he juxtaposes the abstracted notions of Hannah Arendt and Jurgen Habermas with the more dramaturgical conceptualization of public space identified within his own work and that of Clifford Geertz. They do not do so because the narrativization of place and space becomes performatively constitutive of the places themselves.

It is possible to identify a set of relationships operating at the interface of government and civil society that collectively generate racialized landscapes of safety and danger in the governed territories of urban Britain. Crudely this might be summarized as follows. At a banal demographic level specific socio-economic conditions and levels of historical provision generate particular concentrations of young people in specific localities. However, *the wider youth question* (Cohen, 1997a) only emerges from the manner in which problems associated with young people are interpreted through local networks and translated into the policy sphere through particular regimes of governmental knowledge, power, and practice. And this is mediated by specifically local moral panics and debates around community safety and public danger.

At the heart of this process is a communication of generalized perceptions through rumour, gossip, and repute. The place on the political agenda of the youth question is shaped by local fears of the behaviours of young people as much as by the recognition of any 'real' social need. The form of youth-oriented provision is shaped by regimes of funding that reflect the positioning of troubled urban areas as spaces of governmentality. The sites and the projects that are funded reflect the *unsaid* as much as the *said* that emerges from a politicized landscape of institutions and people competing for scarce resources. This becomes even more pronounced in the context of the actions of young people operating within the city spaces territorialized by these governmental regimes.

Repetition, repute and disruption

Although the officially authorized discourse of racial danger and community safety rendered local history in rather similar ways in the Isle of Dogs and Deptford, these concerns were articulated very differently in the geographies of risk shaping local youth and community relations in each place. On the Isle of Dogs, the language of 'rights' and 'fairness' used by white young people and their parents articulated a strong sense of outrage around new housing allocations to Bengali families, while their territorialized fears were focused on the presence of Bengali boys. These themes were taken up and reworked, often in a displaced form, in stories of 'Bengali on Bengali' violence foregrounded by the press and police locally and communicated through dense social networks.

In contrast, in Deptford the official rhetoric of race bore a much closer relationship to the neighbourhood nationalism identified by Back (1995) in an earlier study of the area, which held out the possibility of intercultural identification linked to a strong, officially sanctioned, problematic of neighbourhood harmony. In this context the embedding of 'good vibe' stories in a critical mass of positive urban imagery enabled them to travel across racial divides, although, as we were to discover, there were important instances of discrepancy between these 'authorized' story lines and the plotting of racial encounter on the streets.

The subject of 'youth gangs' was also interpreted differently in the two areas. In Deptford, there was a strong sense that youth gangs, which were predominantly male but sometimes also with small numbers of female members, were distributed as distinct territorial units throughout south London. Gangs had emblematic territorial names like 'Ghetto boys', 'Deptford men', and 'Brixton youth' and at one level they bore the marks of a 'baptismal naming' linked to the popular iconography of black street crime (Kripke, 1980). While talk about gangs was pervasive only three of the young people interviewed had any direct connection to these gangs. In some of the accounts by young white people, these gangs were viewed as a diffuse source of anxiety and threat. At the same time, however, there was a reluctance to racialize these feelings or focus them in some concrete embodiment of the Other, for example, by invoking the figure of 'the black mugger'. At the end of one interview in which a young white girl had talked extensively and quite fearfully about the threat of gangs, she was asked if there was anything else she would like to say. After a short pause she replied, 'Yeah, about the gangs and that. It is true that they are mostly black boys involved but it's not because they're black'.

For the black boys in particular the symbolic nature of gang labels was often in evidence; much of this had to do with claiming male prides of place that might otherwise (for example, in the domestic sphere) be heavily contested. To proclaim 'ghetto boy' status brought with it both a space of preferred identity and a claim to exclusive entitlement. For instance, the area around the Milton Court Estate in New Cross was referred to as 'Ghetto' and as one boy pointed out 'To me now to be from Ghetto is an honour'. Such claims often projected the terms of masculine dominion into wider domains of 'imagined community'. For example, young men would give the numbers of the local telephone boxes as if they were their own public/private home from home. The talk about gangs became a means through which these young people positioned themselves in an alternative map of belonging.

The ludic dimensions of the process could be limiting as well as empowering, particularly for young women. A black girl, complained of being labelled by her male peers:

> Just because you know certain people doesn't automatically make you a part of a gang. Like saying I am like a Peckham girl. I don't think so! I live in Ghetto and the Peckham Boys think I am a Ghetto Girl but because I was born in Peckham the Ghetto Boys think I am a Peckham Girl.

Because baptismal naming is such a powerful rhetorical device for inscribing myths of origin and destiny in local prides of place, it is very difficult to detach these labels from positional attributions once they have initially been made to stick. Nevertheless, while gang territories were acknowledged to be dangerous places it was generally agreed that the risk from gang violence was low unless you are involved directly in that world.

The situation on the Isle of Dogs differed considerably. There was a pervasive sense that 'Asian Gangs' constituted the main threat. Here issues of crime, violence, and risk were strongly racialized and associated with a specific ethnic category: young Bengali men. What became clear in discussing youth violence in the area was the split between black and white youth on the one side and Bengalis on the other. This distinction was kept very much alive in local circuits of rumour and gossip; this, in turn, authorized young people to appeal to adult authority and 'common sense' in validating their iterative statements. This is clear in the following account by a young white woman:

Like even black and whites fight and things like that, but you don't hear of many white people fighting with weapons and things like that, that is the thing. They couldn't just fight with their fist or anything; they have to have their knives and things like that and that is the worst bit. Like my Dad said years ago there wouldn't be weapons, that it would be like getting in a ring and that, and my Dad says they should put them all in a ring and let them just beat each other up, using nothing. But these days they have to use weapons all the time [ ] It just the Asians, the boys act 'flash', sit in their cars as if they own the place.

Here it is the Asians who constitute the threat, who use weapons, fight unfairly, and act 'flash'. This was also present in the accounts of young white boys. We used video walkabouts so some of the young people could make their own films to narrate the spaces of their neighbourhood. One case involved three white boys. As they were walking through Mudchute Farm on the Isle of Dogs the conversation turned to Bengali youth and their lack of mixing with white peers:

David: They all hang out together they never hang out on their own.
Stephen: Like us. We never hang out on our own.
Russell: [referring to Asian youth] The Bangles!
David: The Mukhtas!
Stephen: They re-named Brick Lane. They named it Banglatown – that takes the piss. I wanna go down there and blow 'em all up.
David: We're in England, not fuckin' Bangladesh for fuck's sake. Fair enough if we were in Bangladesh like, they wouldn't name a town [...] – fucking London!
Stephen: I'd get killed, mate, if I went down Banglatown. I'd get killed.

This extract illustrates the way in which the walkabout provided the context for a dialogic, yet highly racialized engagement with processes of baptismal naming: 'Bangles' and 'Mukhtas' are immediately associated with the renaming of Brick Lane, which is then seen to remove the area from both local and national jurisdiction. The boys continue:

Stephen: They're all racist down there; they are, they stare at you 'cos they think it's theirs, their town.
David: You see them bowling along (he starts to rock shoulders, strutting holding his head and looking around mimicking 'the bowl').
Stephen: I guarantee you'd find a blade [knife] on every single one of them.

In the course of the walk, the Isle of Dogs is defined, performatively, as a safe place for whites. It can be prone to the incursion of 'outsiders', but it is the areas outside that are understood as comprising an alien ethnoscape inhabited by violent Bengali boys. The Bengali young men are accused of mirroring the exact forms of masculine embodiment associated with white working-class male street culture. Yet these forms (i.e. acting 'flash' or 'the bowl') are also seen as threatening because they either challenge 'normal' (coded white) territories, or establish exclusionary zones into which whites cannot seemingly venture. In this context the function of the pre-cautionary tale is precisely to map out a potential space of adventure, which is at the same time closed off to empirical exploration by the operation of exclusionary rules and rituals of territoriality.

This imagined geography becomes narrativized around specific incidents. This was particularly true on the Isle of Dogs in the aftermath of one particular case of inter-racial violence where a young man had his head split open by a series of blows to the head with knives and machetes. Here the full potency of the construction of Asian gang violence was brought to effect. As a local youth worker put it a 'fear zone' was created. We have documented this incident elsewhere (see Back and Keith, 1999), but here we want to suggest that this violent incident, along with others, signals an important local shift in both the focus and the forms through which racist constructs are made, narratively, into common sense.

Asian gangs in the contemporary East End and some other urban sites in contemporary Britain are now regarded as the prime sources of risk of urban violence (Alexander, 2001). This has developed to such an extent that it supersedes previous concerns over black male criminality associated with motifs of mugging or civil unrest. It was particularly telling that even local activists from the British National Party claimed 'black and white youth' were equally the victims of police 'double standards' about urban violence, while Asian youth were seen to be getting away with attacks on whites and with violent crime. This shift is not complete and echoes of the 'black mugging' discourse are still registered, but locally these formulations are increasingly faint.

At the same time the communicative genres through which these incidents are plotted, remembered, relayed, and given a racial twist seem to be shifting. In these boys' accounts rumour and precautionary storytelling have increasingly come to predominate over the more verbally elaborated and socially embedded forms of gossip associated with the full blooded (in every sense of the word) cautionary tale. Scare stories about what would be likely to happen if you did venture into Banglatown are now, increasingly, the gist of the white peer group conversation about race, place, and identity, not atrocity stories about what happened to so and so.

Of importance theoretically is that some of these stories are organized spatially and others are organized historically. Time and space are both resources through which particular versions of reality are made visible. We are not ignoring the grim confrontation that accompanies such brute violence. Nor are we making claims about the accuracy of such generalizations, only about their power to shape the official understandings of the racialized landscapes on which policy initiatives will take place.

Historicity and spatiality

Some progressive genres of urban social theory have appealed to a language of the vernacular urban to displace dominant histories and geographies of the local with alternative senses of the past and cartographies of the present (e.g. Hayden, 1995; Sandercock, 1997; Zukin, 1991). In both the areas where we worked dominant narratives of inner urban poverty exist in spaces of the city made rich by traces of alternative histories and geographies. It was in Deptford that Francis Drake landed after circumnavigating the globe, and Walter Raleigh laid down his cloak for his queen. Such local histories also hint at the contested stories of Empire: Drake's landing presaged the Middle Passage, the return of Empire to the homeland in the migrations of the post-war years, the notorious confrontations with the National Front in the late 1970s, the conflicts between local Black communities and the police in the 1970s, and the deaths in the New Cross Fire in 1981. Geographically, the Isle of Dogs is often imagined as the heartland of British racism through the iconic first victory of a fascist councillor in a British local election in 1993. Yet in our research young Bengali women at a local school spoke of the potential freedoms of the Island in contrast to a different kind of surveillance in the more segregated area of Brick Lane and Banglatown.

Progressive appeals by urban theorists to a vernacular talk seek to reinscribe history in place by making visible the stories that have been lost to the landscape of the inner city (Hayden, 1995). This fundamentally redemptive project is similar to work carried out in the UK by groups such as Common Ground who attempt to trace the local histories of particular sites (Clifford, 1996). In such work there is a particular valorization of *the local* through an invocation of the imbedded meanings of *place*. While this may at times be critiqued for romanticizing the local, it also serves as an interesting contrast to other stories that are woven around the same landscapes and landmarks.

It is not simply an observation of ethnographic relativism to point out that such historicities and spatialities are contested and evolving. Our work highlighted the manner in which narratives of time and space were resources, performed through acts of memory- and place-making, that policy elites and young people alike used both to make sense of acting subjects, and to set the stage for further social interaction. Predictably, the research highlighted patterns that were much more nuanced and ambivalent than such a simplistic opposition would suggest. In other parts of the research project, we have explored how narratives of 'harmony discourse' and 'conflict discourse' are drawn on strategically by young people in making sense of their everyday lives. However it is interesting how the dominant notions of Deptford as harmonious and the Isle of Dogs as conflictual echo in the rationalizations and explanations of some of the policy elites.

Most celebrations of public space focus on the manner in which forms of sociability are facilitated by particular kinds of space. Richard Sennett

(1990), in one of his more behaviourist texts, even goes so far as to suggest that in this fashion it would be possible almost to design streets full of life and places full of time. In policy discussions, racism or cosmopolitanism are often imputed to places, as almost essential attributes of certain sites of the inner city landscape. Racism and racialized antagonism are rendered comprehensible through folk geographies that inform and structure the local definition of *the youth question*.

Such attribution replicates in form the manner in which particular sites become known as racist places, with identities with characteristics of their own. In part we are arguing against such behaviourist renditions of the identity of places and in favour of a much closer focus on the processes of narrativisation of 'local culture' in the stories that are told by young people, policy elites, community activists, and others. We want to highlight the predictable simplifications that underscore the deployment of 'harmony' discourse in Deptford and 'racial conflict' discourse in the Isle of Dogs. This is more than just saying that patterns on the ground are complex. Racialized landscapes are places that are constructed as subjects through the techniques of governmental practice that emerge as a cumulative effect of national initiatives, their local realizations and the processes of rumour and gossip through which specific sites are made visible. These cartographies are simultaneously reappropriated in both the alternative stories and what Bourdieu might have described as the *bodily hexis* through which spaces are occupied by a corporeal presence (see Robson, 2000). What our work with policy elites demonstrates is the manner in which places are invoked and reinvented through the interface of practice, regulation, and rumour.

Conclusions

> How do men learn to accept painful surprises and disorder? In that acceptance lies the secret of how purification myths come to seem unreal (Sennett, 1970: 109).

It is important neither to glorify nor demonize what is going on at a local level. An analysis that indiscriminately valorizes the actions of young people potentially celebrates the reproduction of racism and intolerance that assumes ever more complex vectors and creates a diversity of victims. But equally an analysis that allows for the potential of narrative practices to create theatres for dialogue between cultures will run the risk of naturalizing social divisions between genders, ethnicities, and age groups in a sanitized built environment.

While having many reservations about the authoritarian thrust of contemporary social policy, we cannot ignore the grim realities of the racialized violence it seeks to address. The experience of young people in cities is still dominated by the anonymity of irrational and brutal racist attacks, both white on black but also increasingly nuanced by more complex patterns of racialized confrontation. Social policy norms with a behaviourist focus that assume that we can discipline the subjectivities of young people through the creation of new rites of passage, communitarian understandings of social responsibility, and an ever stronger labour market focus on the educational process, may not be a particularly effective way to address issues of racialized differences. A process of socialization that emphasizes the need to accept parental norms and the ethics of labour market competition can play out in some perverse ways when mediated by rhetorics of lineage and of white unemployment both north and south of the River Thames.

If this is the case there is a sense in which an understanding of the public sphere which thinks more contextually about the arenas in which the citizenship of young people is both given and restricted might form a more coherent way forward for thinking about issues of community safety and racial danger in some of the more troubled urban contexts of contemporary Britain. In an important work written over 30 years ago Richard Sennett (1970) drew on Max Weber's distinction between an ethic of responsibility and an ethic of ultimate moral ends. As Sennett (1970: 111) put it 'a responsible act, Weber said, is always impure, always painfully mixed because of diverse motives and desires; an absolute act on the other hand, is a struggle towards purity of desire and act as well as toward a "pure" end'.

In part we are trying to highlight the discrepancy between a political project that normalizes a national agenda at a local level through a set of institutional reforms that we would equate with a set of governmental localizing practices. Such a project attempts to purify the public sphere, suppressing local autonomy in the name of self-government. Potentially it anaesthetizes precisely the sort of surprising settings that Phil Cohen (1997b) has characterized elsewhere as 'the urban uncanny'.

Intercultural dialogue is premised upon transcending particular kinds of subjectivity. It is about escaping, however momentarily, particular forms of disciplined selves and sharing something with a stranger, communicating across boundaries that might only be the product of time and space but are no less powerful for being so. A crude caricature might be that in a Habermasian invocation of the city the exigencies of time and space become obstructions to effective communication. Transcultural dialogue is merely obscured by contingency. Yet perhaps there is something extraordinarily naïve about the manner in which such optimism draws on an implicitly behaviourist understanding of both temporality and spatiality.

There is a sense in which we are created as racialized and gendered subjects in part through the regimes of spatiality and temporality in which we are recognized. This might point to both an understanding of temporality and spatiality as constitutive of subjectivity rather than just as arenas in which subjectivity is performed. It might also suggest that we need to pay a closer attention to the times and places at which sublime moments of identification occur that challenge the taken for granted self and look to both an aesthetic *and* an ethics of the production of public space.

Notes

1 Our research was funded by the Economic and Social Research Council (Project number: R000236301) and carried out by a research team that also included Phil Cohen, Tim Lucas, Tahmina Maula, Sarah Newlands and Lande Pratt.
2 Interesting mappings of the local areas of safety and danger by young white students at George Green School identified a barrier of 'Bengali danger' across the north of the Isle of Dogs that corresponded closely with these estates.

References

Alexander, C. (2001) *The Asian Gang*. London: Berg.

Back, L. (1995) *New Ethnicities and Urban Culture: Racisms and Multiculture in Young Lives*. London: UCL Press.

Back, L. and Keith, M. (1999) 'Rights and wrongs: youth, community and narratives of racial violence', in P. Cohen (ed.) *New Ethnicities, Old Racisms*. London: Zed Books. pp. 131–62.

Clifford, S. (1996) 'How many common streams? Places, cultures and local distinctiveness', in R. Jaijee, and K. Thomas (eds), *Getting in Touch with the Thames*. London: London Rivers Association. pp. 16–20.

Cohen, P. (1997a) *Rethinking the Youth Question*. London: MacMillan.

Cohen, P. (1997b) 'Out of the melting pot into the fire next time: imagining the East End as city, body, text', in S. Westwood and J. Williams (eds), *Imagining Cities*. London: Routledge. pp. 73–86.

DoETR (1997) *Building Partnerships for Prosperity*. London: HMSO (Cm 3814).

Driver, S. and Martell, L. (1997) 'New Labour's communitarianisms', *Critical Social Policy*, 52: 27–46.

Giddens, A. (1998) *The Third Way*. Cambridge: Polity.

Gordon, C. (1991) 'Governmental rationality: an introduction', in G. Burchell, C. Gordon and P. Miller (eds), *The Foucault Effect: Studies in Governmentality*. London: Harvester Wheatshaft. pp. 1–53.

Hayden, D. (1995) *The Power of Place: Urban Landscapes as Public History*. Cambridge, MA: MIT Press.

Johnston, L. (2000) *Policing Britain: Risk, Security and Governance*. Harlow: Pearson Education.

Keith, M. (1995) 'Ethnic entrepreneurs and street rebels: looking inside the inner city', in S. Pile and N. Thrift (eds), *Mapping the Subject*. London: Routledge. pp. 355–71.

Kripke, S. (1980) *Naming and Necessity*. Oxford: Blackwell.

Lees, L. (2003) The ambivalence of diversity and the politics of urban renaissance: the case of youth in downtown Portland, Maine, *International Journal of Urban and Regional Research*, 27: 613–34.

Mitchell, D. (2003) *The Right to the City: Social Justice and the Fight for Public Space*. New York: Guilford.

Mulgan, G. (1994) *Politics in an Anti-Political Age*. Oxford: Polity.

Robson, G. (2000) *No-one Likes Us, We Don't Care: The Myth and Reality of Millwall Fandom*. London: Berg.

Rose, N. (1989) *Governing the Soul: The Shaping of the Private Self*. London: Routledge.

Rose, N. (1999) *Powers of Freedom: Reframing Political Thought*. Cambridge: Cambridge University Press.

Sandercock, L. (1997) *Towards Cosmopolis: Planning for Multicultural Cities*. New York: Wiley.

Sennett, R. (1970) *The Uses of Disorder: Personal Identity and City Life*. New York: Norton.

Sennett, R. (1990) *The Conscience of the Eye*. New York: Norton.

Sennett, R. (2000) 'Reflections on the public realm,' in G. Bridge and S. Watson (eds), *A Companion to the City*. Oxford: Blackwell. pp. 380–8.

Smith, N. (1996) *The New Urban Frontier: Gentrification and the Revanchist City*. London: Routledge.

Zukin, S. (1991) *Landscapes of Power: From Detroit to Disney World*. Berkeley, CA: University of California Press.

5 The Emancipatory Community? place, politics and collective action in cities

James DeFilippis and Peter North

even in big cities people continue to act collectively at times on the basis of common territory: the people of a neighborhood resist urban renewal, white homeowners band together to resist black newcomers, disputes over the operation of schools bring geographical groupings clearly into view ... their very existence identifies the need for a better understanding of the conditions under which collective action on a territorial basis occurs. (Charles Tilly, 1974: 212)

At the heart of my beliefs is the idea of community. I don't just mean the local villages, towns and cities in which we live. I mean that our fulfilment as individuals lies in a decent society of others. My argument ... is that the renewal of community is the answer to the challenges of a changing world. (Tony Blair quoted in Levitas, 2000: 189)

We begin with Tilly's (1974: 212) unambiguously affirmative answer to the question 'Do Communities Act?' because we believe that collective action remains a major, if sometimes dismissed or overlooked, political component of urban life in Western cities. Furthermore, we argue that it is most often in the shared territorial spaces that are constructed to be communities where the city's celebrated ability to allow for the formation of collective political identities and consciousness is realized. This is neither to assert nor deny the normative desirability of the idea and ideal of community, although these are issues we will discuss in this chapter, but rather to recognize that it is a conceptual framework that is often employed by people and organizations in urban areas.

The second quotation from British Prime Minister Tony Blair highlights the parallel political reality that while the *idea* of community is used to mobilize people to act collectively, the *ideal* of community is increasingly invoked in Anglo-American politics. This is particularly true in the rhetoric and policies of the New Labour government in power in the United Kingdom since 1997. The invocation of community by the Blair government could be easily dismissed as a cynical political exercise designed to put a benign face on the process of state privatization – and there would be no small amount of truth in that depiction (see North, 2000) – but it is also more than this, for political communitarianism has become a core component of New Labour philosophy. Respectively, the two introductory quotes represent the two principal halves of the notion of community: the extent of its literal, empirical existence and its normative ideal – or, as Joseph Gusfield (1975) put it, its semantic and poetic meanings. In confronting these two halves of community, we are foregrounding the processes that occur in the meeting (or collision) of the semantic and poetic meanings of community.

In this chapter we begin by asking why collective action around shared territory, in the name of community, is an almost inherent component of urban life. We then discuss the political debates surrounding the ideal of community, particularly in the context of New Labour and its emphasis on it. The bulk of the chapter, however, will draw on our experiences as activists and participant-observation researchers in the regeneration processes in the Elephant & Castle area of south central London.[1] We have consciously tried to limit the extent to which our discussion of community and the potential for emancipation is an abstract one. Instead, our discussion focuses on the emancipatory potential of community based organising in the contemporary world of British politics.

Place, community and collective action

Community is one of the most ideologically loaded terms in the English language. Rather than define community here, which is well beyond the scope of this chapter, we accept that there are two broad sets of meanings attached to the word. The first is some kind of self-defined group that, in the words of John Agnew (1989: 13) shares 'a morally valued way of life'. These are what could be called communities of interest, and in an urban environment this can mean people who might not live together in a given space, but nevertheless congregate away from their homes to pursue their shared interests. The second is geographically defined (even if that definition is porous and mutable) as 'social relations in a discrete geographic setting' (Agnew, 1989: 13). These two senses of community are clearly related, as shared communities of interest and affiliation often emerge from place-based social relations. But they can also become confused in discussions of community, especially when drawn into divisive efforts to define who is or is not the 'authentic' community: Is it 'residents'? Members of faith groups who live elsewhere but worship in a shared space? Asylum seekers or immigrant communities? Members of voluntary or community groups that are dependent on funding from state agencies? These are questions that can tear community groups apart. Even in the most segregated cities, differences and conflict are integrally part of intra-community social relations. While helpful for analysis and essential for forging affiliations that work with and through difference, we would also agree with David Harvey (1996) that fixation with these differences can be a barrier to the construction of local urban community groups who are able to act effectively in their collective self-interest.

While the importance of place in social relations has now long been recognized by geographers, they have not considered fully how these interactions influence political consciousness and collective actions at the geographic scale of the community. The relationships between place and collective action have most explicitly been addressed in the context of identity-formation and identity politics (see, for instance, Keith and Pile, 1993; Miller, 2000; Pile and Keith, 1997), often, unfortunately, at the expense of elaborating the importance of urban structures (the nature of and use of buildings and urban spaces) and collective consumption (housing, community facilities and the like). Perhaps the limited attention given these issues is a function of the continued rejection of the structuralist analyses of the 1970s, in which they were foregrounded. While it is unquestionably true that identity (trans)forming relationships are often interwoven with place, and therefore politically and intellectually important to understand, discounting structural issues of, for example, housing access, risks overlooking some of the basics of collective action in urban space. As Joseph Kling and Prudence Posner argue:

> Any activist knows that in the United States probably the easiest issue to mobilize people around is the protection of their property rights. Second easiest is the demand that 'the community' participate in decisions that affect the life situation (e.g. property value, child raising and education, shopping, traffic patterns …). (1990: 36)

Their first mobilizing rationale has a definite class and owner-occupier bias, but if we substitute 'home' for 'property rights' then it would hold much more broadly. Thus we would suggest that at the heart of collective action in the community is defence of the *home* and the means of *collective reproduction* (for housing, education, health care, etc. are not simply realms of consumption but spheres of active, if often unwaged, labour). This is because place-based communities are the sites of residences and the relationships and activities associated with them. For this reason some theorists (e.g. Smith, 1993) have defined the geographic scale of the community as the scale of social reproduction. On some irreducible level, therefore, it is only logical that issues surrounding the home should be at the fore of community-based collective action. The questions then become: how do relations of domestic property lead to collective action in urban space? What is the emancipatory potential of such action?

In one of the most useful, if over-looked discussions of these issues, John Davis (1991) worked through a framework for understanding how place and territory interact with people's relations to their domestic property to create the basis for collective action in urban space. In his work the functional and tenurial relationships (that is, relations of control and ownership) that people have with their domestic property shape and define the potential for collective action in urban space. His understanding of how relations of domestic property can lead to radical political mobilization, however, is more than just taking Karl Marx's transformation from a 'class in itself' to a 'class for itself' and applying it to a neo-Webberian understanding of housing classes. Instead, the model is a three-staged process that begins with *collective consciousness*, which is the recognition of shared interests and property relations (as, for instance, council housing residents). It then moves to *conflict consciousness*, which recognizes not only shared interests, but also how those interests differ from, and are in conflict with, other property interests (such as the local council or a would-be developer). And it progresses to *radical consciousness*, which is the realization that the current structures and relationships governing property are inherently unjust.

We recognize that this schema is very mechanistic, and it is not meant to be a definitive guide to radical community-based collective action. We also readily acknowledge that all too often the process works in reverse. Community groups that emerge out of a radical critique of society can find themselves transformed over time into not-for-profit housing developers and landlords (see DeFilippis, 2003: ch. 2). Similarly, the process doesn't always 'progress' and groups formed to improve local housing conditions remain as such. This is particularly true with middle class homeowner associations, which can be very quick to mobilise, but have little interest in any radical reconstruction of property relations. For instance the most important and successful American social movement of the last quarter century – the anti-tax movement – emerged as a homeowner movement in California and spread from there, bringing Ronald Reagan to power. In short, we do not assume here that consciousness raising and transformation, and changes in the character and goals of groups involved in collective action, are either natural or inevitable developments. Nor do we assume, as Manuel Castells (1983) did, that collective action around issues of domestic property are necessarily progressive or radical. They are clearly not – and the movement towards a radical social critique is not in any way an unproblematic process. But these concerns notwithstanding, the schema proposed by Davis (1991) remains a politically and intellectually useful one, and one which is largely borne out in the case of the Elephant and Castle.

The ideal of community and emancipation

But if we have, briefly, sketched out a framework for collective action in urban communities, we have not dealt with the inherently political question of whether or not the community should be the focus of collective action. The notion of community as the basis for collective action has been challenged primarily from three different directions, Marxists, feminists, and post-structuralists.[2]

The Marxian critique of community as a basis for collective action and social change is that communities are not the realms in which the dominant frameworks of power and exploitation in society are produced and reproduced. If class relations are the foundation of social relations then to organize in 'the community,' even if communities are the spatial expressions of people's class positions, is largely to treat the symptom rather than the cause. As Harvey (1981: 115) states, 'This leads us to the notion of displaced class struggle, by which I mean class struggle which has its origin in the work process but which ramifies and reverberates throughout all aspects of the system of relations which capitalism establishes'. To be sure, there have been more nuanced discussions of collective action in the community from Marxists, and Ira Katznelson's (1981) work stands out in this regard. But even in Katznelson's work, there is a politically destructive (to the left) schism between class and community in urban politics. And to Katznelson, this gulf is largely because community politics have slipped far too easily into the pre-figured 'trenches' of community issues and lost sight of class relations and conflict. And while some components of the labour movement in the US and the UK have become more involved in community organizing efforts, and bridges have been built between community groups and organized labour – most notably in the living wage movement which has spread throughout the US and even made it to the East End of London (see Littman and Wills, 2002) – the political gaps are still wide and difficult to overcome.

Feminist and post-structuralist critiques of community have focused on its oppressive effects on individuals in general, and particularly on those different from, or outside of, the dominant social group within the community, and so, despite important differences within and between them, can be conveniently discussed together. In its abstract expression, this critique borrows from the debates between liberals and communitarians, while agreeing with neither. It was perhaps most clearly and thoughtfully articulated by Iris Marion Young when she stated:

> The ideal of community, I suggest, validates and reinforces the fear and aversion some social groups exhibit toward others. If community is a positive norm, that is, if existing together with others in relations of mutual understanding and reciprocity is the goal, then it is understandable that we exclude and avoid those with whom we do not or cannot identify. (1990: 235)

Similar arguments have long been made by feminists (see, for instance, Friedan 1984). These critics are surely right that modern and post-modern cities that can liberate individuals from the oppressive conformity of small town/rural or suburban life. It is precisely these kinds of oppressive conformity that have led authors like Young (1990) and Elizabeth Wilson (1992) to celebrate the emancipatory potential of the diversity and potential anonymity offered in urban space. And it is this celebration of anonymity that leads such authors to significantly challenge the normative ideal of communities.

The debates surrounding the semantic and poetic meanings of community, however, have largely been unheeded in contemporary British politics. Community is, almost literally, the 'Third Way' between the society-centred framework of Old Labour and the individualist-centred perspective of The New Right. The notion of community invoked by New Labour follows Amitai Etzioni (1993, 1996) and borrows heavily from Robert Putnam (2000). Like these intellectual models it is almost completely devoid of class and class conflict and pays little heed to the potentially repressive and intolerant character of communities, which troubles so many critics (e.g. Lees, 2003) . Tellingly, in his discussion of communitarianism and contemporary social policy, Adrian Little (2002, ch. 6) discusses only one particular policy arena in any great depth: crime control. It is thus a simultaneously controlling and conflict-free understanding of community informing the dominant political communitarianism in Anglo-American politics.

The intellectual debates around the problematic nature of the ideal of community are therefore particularly important in refuting the simple-minded political communitarianism of public figures like Tony Blair. At the same time, however, they can also potentially undermine the efforts of low-income urban communities to organize collectively in pursuit of their goals. This potentially emancipatory aspect of urban life – the ability to organize collectively – has long been recognized by social theorists, and it was certainly part of Marx's understanding of the potential of urban politics. All this, ultimately, is what leads Judith Garber (1995: 37) to observe, 'as an abstraction, local community is deeply problematic; in practice, it may actually serve women more often than we think.' We agree with Garber's assessment, even if we are not as sanguine as we would like to be about the outcomes of community-based struggles against larger-scale structures and institutions with greater political capacity. For us, the interesting questions are not: are there communities? There are. Or, can they act collectively? They can, as work on urban social movements shows (e.g. Castells, 1983; Lowe, 1986). Rather we want to ask: when they do emerge? How are collectively organized community-based agents constructed? How do they act and with what success?

With this background in mind we now explore these issues in a case study of conflict over the regeneration of the Elephant and Castle in South London.

Elephant Links

Elephant Links is an urban regeneration programme centred on the Elephant and Castle in south east central London. It's the missing part of central

London: typically maps of central London show neighbouring Westminster, the West End, and the City, while the Elephant is obscured by the legend. Yet it is less than a mile from each of those quarters and close to the revitalized South Bank, which now attracts tourists by the legion to its Tate Modern, 'Wobbly' (Millennium) Bridge, the London Eye, HMS Belfast, and the GLA Building at Tower Bridge. It has a locally popular, yet run down shopping centre, which, incidentally, was the UK's first covered shopping centre. The Elephant is one of London's major traffic junctions for road, rail and bus and suffers from high levels of congestion. Finally, the Elephant is home to gangsters, MPs, architects, journalists, and some of the highest levels of social exclusion in the city. It therefore encapsulates both the emancipatory potential (its density, diversity and plurality) and the oppression of urban life (its concentrations of poverty). Although such concentrations in-and-of-themselves do not automatically have emancipatory potential, they can, and in this case do, provide the space for the collective organizing necessary to realize social change in urban space.

Given its prime location close to central London, the Elephant is also an area ripe for redevelopment. In 1999, the London Borough of Southwark led a partnership of local people, businesses, voluntary organizations, and other public sector organizations that won a Single Regeneration Budget (SRB) funding bid from the central government. A private sector partner in development was chosen the following year to carry forward the plans for regeneration in the area. The Elephant Links programme was about the transformation of urban space and the communities living there. It was also about the age-old question: who would benefit from this transformation? In what follows we discuss how community-based activists at the Elephant used discourses of 'community' in an emancipatory project that aimed to ensure that regeneration would benefit local working-class people, rather than just becoming another example of gentrification.

The emancipatory potential of community consultation

British urban policy – in 1999 anyway – included a requirement that local residents affected by regeneration should be fully included in discussions about future plans for the areas they inhabit. Theorists from Sherry Arnstein (1969) have debated the extent that consultation has emancipatory potential. And it was here, literally at the beginning of the regeneration process, that conflict over space erupted. To inform its bid, the council convened a Residents' Regeneration Group (RRG) made up of local people who met for 18 months and developed their own 'principles for effective regeneration': the need to improve – and replace any lost – council-owned housing,[3] environmental improvements to local green spaces, and improved community facilities. But these considerations were not at the heart of the council's regeneration agenda. In the eyes of the Labour-led Southwark council, local residents would be consulted, but they would not be in charge of the regeneration process. The main themes of the bid for SRB funding were worked up by council officers and presented to other partners somewhat late in the day in what appeared as a fait accompli. The bid did not address major community issues around housing and the need for community facilities. The bulk of expenditure would go to provide environmental and transport works to facilitate physical development and the council's flagship social inclusion programme, which looked to help unemployed people to find work (a classic New Labour 'workfare initiative'). Thus the bid focused on top down New Labour-friendly social inclusion policies and the facilitation of land assembly for the development, rather than the community-generated programmes developed by the RRG.

Residents had to fight to get the regeneration programme to address their concerns. They threatened that unless their views were included they would not support the SRB bid, thereby jeopardizing the council's likelihood of getting the £25 million it was seeking to win from central government. Council officers responded that the community's agenda would not adequately meet the government's criteria for funding. Thus while residents were free to put forward their views, the extent to which the process was open enough for residents to pursue their own visions for the Elephant and Castle was limited by the need to meet priorities generated centrally. At their root, then, conflicts, struggles and barriers to emancipatory community organizing in the Elephant stem from this set of power relations governing the regeneration process. While other partners in regeneration saw local residents as objects of, rather than actors in regeneration, residents were able to use government rhetoric about the importance of participation to insist that they be treated as full decision-making partners – not consultees.

Community activists resolved to make Elephant Links a partnership led by local residents. They were concerned that the council was working to a hidden 'social cleansing' agenda. In an infamous remark, Southwark's then Director of Regeneration, Fred Manson, argued that 'We need to have a wider range of people living in the borough … social housing generates people on low incomes coming in and that generates poor school performances, middle class people stay away' (quoted in Wehner, 2002). Southwark, it was argued, suffered from having too many of the 'wrong sort' of residents: socially excluded people disadvantaged not by exclusionary labour market processes in a global city, but by 'low' aspirations and low social capital that they passed on to their children. The council's answer was managed but inclusive gentrification to bring in more wealthy residents with higher levels of social capital and labour market involvement and paying higher levels of local tax, which could be used to benefit local residents (provided they were not displaced in the process).

Despite these very real problems of social exclusion, community representatives refused to scale back on their vision for the Elephant. They wanted to be involved in the SRB and make it work for local residents. For some, their communities needed to be defended from gentrification, and they had a right to a voice. For others, the broad aims of the SRB programme were loose enough that community benefits could be secured in the future, and residents would be able to influence the redevelopment of the shopping centre and their estates. This is what is so interesting about the Elephant. Even the more radical community-based voices in the Elephant wanted to be constructive. They did not feel that the changes proposed were all bad or that the Elephant was fine as it was. They took the SRB rhetoric of partnership seriously and wanted to be involved in working to improve the plans rather than shouting from the sidelines. As we shall see, the tragedy of the Elephant is that Southwark's actions pushed these voices into an oppositional stance that they had tried so hard to avoid. But on the way, through engagement, local residents also won a number of victories.

Winning resources for an emancipatory project – the Elephant Links Community Forum

Partly as a result of this activism, a Community Forum was established by the Elephant Links partnership through which local tenants and residents could get involved in the SRB. The Resident's Regeneration Group (RRG) joined the forum, believing that it could be an effective vehicle for promoting community interests. But very quickly it became clear that the Forum would need its own staff and resources so that local people could play an equal part alongside the better resourced private and public sector participants. After considerable conflict with the council over how quickly staff and technical support could be hired, the Forum activists won funding from the SRB to hire staff and create an office. An office was opened in the shopping centre, which became a focus for organizing around local residents' views of the future of the Elephant. Residents had won their own voice, their own resources, and a space from which they could develop further their ideas for the future of the Elephant.

Community action through partnership mechanisms

British urban regeneration policy assumes that members of the partnership boards disbursing government regeneration monies act as individuals whose first responsibility is to ensure that the expenditure of public funds is accountable to the government and consistent both with National Audit Office rules and the aims of the SRB. They are thus managers of money, not local representatives. The Forum's representatives were wholeheartedly committed to ensuring that grants were properly spent, but they believed that this could best be done by ensuring that projects benefited local residents. As they were only five out of the 21 members of the Partnership Board, they felt they needed to act collectively at the board. They met before meetings, made sure they understood the issues, debated, and (eventually) agreed a common position. As one Community Forum representative put it 'we are more together. We disagree amongst ourselves, but we are a united front, perhaps we are more interested. I think we were allowed to dominate, really'. They effectively defended their views of what the Elephant should look like.

Acting through the Partnership Board residents sought to influence decisions on their main priorities – housing and the nature of the built environment that would emerge from the regeneration process. This was a constant battle. The council accepted community participation in discussions about the physical development, but argued for consultation rather than a decision-making partnership. Papers described this or that decision that the council would make without reference to the Partnership Board. The council argued that the physical development was a public-private partnership between the council and a development partner, while the Forum argued for holistic regeneration managed through the Elephant Links Partnership Board.

The impasse was solved when, as a result of community action, the council granted residents a key voice in the selection of the development partner. The Community Forum helped decide the criteria for the technical appraisal, and after the Community Forum, the partnership board and the council independently reviewed each of the three competing consortia on offer, a consensus was achieved over one development partner. However, more radical voices within the Community Forum were concerned that while a creditable amount of consultation and information sharing had been developed through the selection process, it did not amount to shared power or decision-making authority. The council consulted, but made the final decision, and more importantly managed information flows between the developer and the community. So in another successful piece of community organizing, the Community Forum got the board to agree to a more robust tripartite structure in which the board, the council, and the developer would work as partners, with significant levels of independent technical support from the SRB, to manage the master-planning process.

The council responded by trying to limit the role of the tripartite body. Meetings took place between the developer and the council without the community representation. Despite repeated efforts by the council to control the development process, the Forum held out for full equality of decision-making, which was finally agreed in February 2001. After these battles, the community seemed to be in a strong position to influence housing and the nature of the built environment as plans for the redevelopment of the Elephant began to emerge. But this had been achieved at some cost. Conflict had become embedded in board meetings. The Community Forum took up considerable time arguing its case and walked out when progress was not made to its satisfaction. An article in *Property Week* painted the following picture: 'Once official proceedings begin there is not a spare seat in the House. The Community Forum lines up on one side and local councillors and interested parties on the other in a spaghetti western-style stand off' (Creasey, 2001: 46). The other partners' sympathy for the Forum was, to say the least, beginning to wear thin. They felt that the community side was not paying sufficient attention to the core task of managing the SRB programme, which was considerably behind schedule.

The Forum's perception was that the best way to solve endemic conflict was to change the structure of Elephant Links so as to be community-led and to set up a company with its own staff and budget. This company would then grow into a development trust, which would act as the vehicle into which the community benefits from the development would be vested. Another innovation suggested was a Community Land Trust (CLT), which would own the new social housing that came out of the development. A CLT is a particularly radical and innovative form of property ownership in which the land is taken out of the market, the housing units on it are rendered permanently affordable, and community control is written into its governance structure (DeFilippis, 2002). From the community side, this was seen as a more innovative way of securing community control of social housing than either council housing or registered social housing, which was seen simply as a vehicle for privatization. But here, again, the council procrastinated. Asserting itself, the Community Forum asked for feasibility work for a CDT, which council officers felt was premature. Rather than taking instructions from the Board, they resisted a community land trust proposal in favour of a registered social landlord (or housing association) form of housing ownership.

Overall, then, the community had won a significant role in decision-making and seemed well placed to influence the development of the Elephant so as to meet the needs of the existing working-class communities. But community representatives were also criticized by other partners for the way they influenced board policy through effective organizing and caucusing before meetings, rather than through negotiating and taking other partners with them.

Things fall apart: the end the Forum

The pressure began to tell. Forum members began to fall out with each other, as radical and more conciliatory voices clashed. Representatives from the Heygate Estate, the largest council estate in the Elephant, left the Community Forum because they felt that their core concerns were not represented adequately in an organization that, as a consequence of the need to fulfil SRB outputs, had significantly widened its membership beyond housing to include some 63 local community-based organizations. Their decision seriously weakened the Forum's claim to represent the local community.

The Forum was in a position to clog up the workings of the partnership if it was not listened to, and that gave it a power it was ready to use. The partnership entered a situation in which neither side could impose its will on the other. This was unsustainable, and the outcome was perhaps predictable. The council organized against these 'unruly subjects', and gained control over what, for it, had become an unmanageable process. After two tied ballots, the Community Forum representative chairing the partnership was deposed by a local business person. Within minutes of his election he began to run the board in what was seen by the community as an exclusionary and authoritarian manner that ditched all pretences towards local democracy in favour of the management of the SRB. The chair and the project director would henceforth set the board agenda. An attempt was made to exclude the Community Forum's director from the partnership table. Community voices from the floor would no longer be called to speak. After the withdrawal of Heygate from the Community Forum, the chair also attempted to reduce its representation on the board. The Community Forum responded that local residents should now form a majority of the Partnership Board. As neither the new chair nor the Community Forum would back down, the result was all out war at board meetings.

The council then came after the Community Forum's resources. In an evaluation, the Forum was accused of 'poor judgement' in prioritising development rather than what were called its core objectives. It was accused of failing to recruit enough volunteers, even though it had grown from 17 to 63 participating organisations. The council then refused to pay the Forum's grant, and its staff were issued with redundancy notices.

The community representatives were 'othered' by council officers who held the partnership to ransom. They charged that the Forum was unduly influenced by its director and a small group of politically-motivated activists opposed to the development. The council systematically undermined the Forum's claim to legitimacy as a community voice by claiming the 'community' was not a unitary actor. It ignored the considerable work the Forum had done to build a federal structure that would ensure democratic legitimacy and that the diverse communities at the Elephant were properly represented. The Forum was put under scrutiny and its identity was unpacked in ways designed to neutralize it, while the legitimacy of the public and private sectors went unchallenged.

At the same time, the council voted to terminate the agreement with the selected private developers. It also sought a new housing-based development on the Heygate involving a consortium of registered social landlords. At this point, it became clear that the council had, for some time, been negotiating separately with housing associations on proposals for the Heygate site, reversing the Community Forum's earlier victory in winning full replacement of council housing in the new development (MacDonald, 2002).

The Community Forum, by this time without its staff, challenged these decisions at what was to be the last meeting of the Elephant Links Board in June 2002. The chair attempted to rule the resolution out of order, and a near riot erupted as members of the local community vented their collective spleen at what they saw as the combined and multiple injustices of the SRB process. The community chanted out the chair's voice, while the Forum then served an injunction on the board demanding that their funds be restored. After unsuccessful attempts to clear the room, the meeting was suspended and the next day the chair decided that the board was no longer competent to administer the SRB.

The council then moved quickly. It raised the stakes and called for the Forum to turn over all its documents. Rather than attempting to pursue mediation, Southwark went to court and gained an injunction, which was served on Forum management committee members at home, requiring them to hand over all documents, freezing the Forum's bank account, and prohibiting use of any of the Forum's assets. Under pressure, the Forum management committee split and narrowly voted to provide the council with the information it needed. The radicals wanted to continue the battle, and on losing the vote felt that they no longer wished to be on the committee. They resigned.

Elephant Links was reconstituted by the Council without significant community involvement, and in 2004, new plans for the regeneration of the Elephant were unveiled which had been drawn up without community consultation.

Conclusions

In this chapter we have presented an in-depth set of participant observations about the reality of community-based organizing in contemporary British urban politics. There are significant lessons about the emancipatory potential of community-based organizing to be taken from the experiences of the Elephant and Castle, where a federation of community-based organizations – the Community Forum – refused to play by the subsidiary role regeneration rules assigned it. The obvious point is that New Labour's rhetoric of community has clearly not been matched in its actual practice. Or, rather, the rhetoric of community has been put into practice, but because the ideal (or poetic meaning) of community being put into practice denies difference and conflict within communities, it yielded a situation of paralysis. Both as an idea and an ideal, community need not erase difference, but any conception of community that ignores conflict and difference will inevitably struggle if differences and conflict 'crash the party' – as they are so often apt to do. In this sense, the emancipatory potential of community-based collective action in urban space collides with attempts to instrumentally impose an ideal of community in which there is no space for collective action.

None of this is particularly striking, but the extent to which community residents and its principal organizational arm, the Community Forum fought with the Labour controlled Southwark Council is remarkable, to say the least. The Forum – or rather the activists at its heart – were consciously able to weld a collective actor from a disparate range of community-based organizations based in an immigrant reception area geographically located one mile from the Houses of Parliament and the City of London. This location matters. At the heart of a global city, population density, geographical proximity, and the palimpsest of layered traditions of urban political action formed a dense and rich sedimented network of information, advice, support, and resources that community activists could call upon.[4] This network ebbs and flows and can be reactivated in the most unlikely of settings. In creating the Forum, the council brought this hitherto hidden network together, and the activists then rebelled against the subsidiary position assigned for it.

These activists then used the rhetoric of SRB to pursue their own vision of the city. SRB rhetoric promoted the idea that effective city management requires the involvement of local people and their agreement to plans – and local residents took this rhetoric at its word and insisted on a voice. Its rhetoric proclaimed that local people should be supported so that they have an equal voice, and again residents took the rhetoric at face value, and won significant resources that they controlled. Activists worked through an urban policy framework that facilitated identity formation within a disparate community, and created a collective actor able to pursue an emancipatory agenda for the city built around the importance of community facilities, green spaces, and social inclusion policies that met the needs of local residents. Urban policy, they argued, should be attentive to the cultural needs of the diverse residents in the Elephant.

The partnership structure formed a channel through which these arguments could be made, and acted as a mobilizing process drawing residents together into a space where they could develop their ideas and form an organisation able to represent them. But it also structured the protest in ways that were not always fruitful. Residents became more and more conflict-oriented in their interactions with the council as the process dragged on. This increasing radicalization was a result of the shared resident interests (based on their common housing tenure status) coming into direct conflict with other property interests. The push for a CDT, and then a CLT are emblematic of this radicalization. But at the same time, knowledge of bureaucratic regeneration procedures and administrative codes and an ability to argue in bureaucratic meetings like a partnership board were both necessary and scarce skills (even if their urban location meant that there was plenty of advice to hand). The mode of organization was rather elitist, emphasizing those with the necessary skills while other members of the community, if they knew about the partnership at all, were simply a stage army who watched as community leaders and the council jousted. Those with skills did not pass them on, perhaps because they did not have the time, but also because they saw themselves as acting *for* their constituents rather than as facilitators of community action from below. There were too few public manifestations of community support for their leaders, and thus their credibility could be challenged, as we have seen, through the systematic undermining of their capacity to speak for a unitary and coherent actor called 'community.'

Finally, there are significant questions that need to be asked about the emancipatory potential of community-based organizing in conflict with the state. This is hardly new, and the experience in the Elephant bears a striking resemblance to the American 'Community Action Agencies' of the 1960s, for which the rhetoric was of 'maximum feasible participation'. They found that the looseness of 'community' was eventually unravelled as radical and conciliatory voices clashed, and newer members of the community were 'othered' by more long-standing elements. In the Elephant traditional housing-based activists felt that as the Forum grew, their housing concerns were diluted as the Forum became dominated by minority ethnic and faith groups. By contrast others felt that housing interests were dominated by older members of the white community and that as it expanded the Forum became more representative of the ethnic diversity of communities at the Elephant. The problem with 'community' is that the representativeness and authenticity of community-based activists can be challenged by opponents unless they pay close attention to grassroots organization and democracy, and in particular are comfortable, and skilled in, working with and through difference. In short, the dilemma of community as an emancipatory category is that its poetic meaning is often simultaneously supportive and disruptive of its semantic meaning.

Notes

1   For one of us this involvement was intensive and occurred over a three-year period as a resident in the Elephant, a member of Elephant Links Community Forum, and an academic in a nearby institution. For the other, the participation was much more limited, primarily serving as a consultant to the Community Forum on occasions. We would like to thank the members of the Forum with whom we worked for the ideas and experiences that we report and comment on in this chapter. Especial thanks are due to Richard Lee, Al-Issa Munu, Anne Keane, Ted Bowman and Celia Cronin, among others from the Forum, and to Julia Brandreth and Karen O'Toole from Elephant Defend Council Housing. They inspired many of the ideas expressed in this chapter, although any mistakes and omissions are obviously the responsibility of the authors.

2   This is not to deny the importance of the long-standing debate between communitarians and liberals. But with their starting point as the primacy of the individual in social life and social theory, liberals have always struggled to make sense of collective action and accordingly have largely chosen not to debate on this terrain.

3 A key element in any emancipatory project for working-class residents in British cities is the defense of local authority owned social housing (called council housing) with secure tenancies and rent control. See the website of the campaigning organization 'Defend Council Housing', www.defendcouncilhousing.org.uk for more information.

4 Many of the key actors at this second 'Battle of Bermondsey' feature in Peter Tatchell's (1983) discussion of his attempt to become the area's socialist MP.

References

Agnew, J. (1989) 'The devaluation of place in social science' in J. Agnew and J. Duncan (eds), *The Power of Place*. Boston, MA: Unwin Hyman. pp. 9–29.

Arnstein, S. (1969) 'A ladder of participation in the USA', *Journal of the Institute of American Planners*, 35: 216–24.

Castells, M. (1983) *The City and the Grassroots*. Berkeley, CA: University of California Press.

Creasey, S. (2001) 'Power to the people', *Property Week*, 30 March: 45–8.

Davis, J. (1991) *Contested Ground: Collective Action and the Urban Neighborhood*. Ithaca, NY: Cornell University Press.

DeFilippis, J. (2002) 'Equity vs. equity: community control of land and housing in the United States', *Local Economy*, 17: 149–53.

DeFilippis, J. (2003) *A Voyage to Lilliput: Collective Ownership and Local Control in the Global Economy*. New York: Routledge.

Etzioni, A. (1993) *The Spirit of Community: Rights, Responsibilities and the Communitarian Agenda*. New York: Crown.

Etzioni, A. (1996) *The New Golden Rule: Community and Morality in a Democratic Society*. New York: Basic Books.

Friedan, B. (1984) *The Feminine Mystique*. New York: Dell Publishing Co.

Garber, J. (1995) 'Defining feminist community: place, choice, and the urban politics of difference', in J. Garber and R. Turner (eds), *Gender in Urban Research*. Thousand Oaks, CA: Sage. pp. 24–48.

Gusfield, J. (1975) *Community: A Critical Response*. New York: Harper & Row.

Harvey, D. (1981) 'The urban process under capitalism: a framework for analysis', in M. Dear and A. Scott (eds), *Urbanization and Urban Planning in Capitalist Societies*. New York: Metheun. pp. 91–122.

Harvey, D. (1996) *Justice, Nature and the Geography of Difference*. Oxford: Blackwell.

Katznelson, I. (1981) *City Trenches*. New York: Pantheon Books.

Keith, M. and Pile, S. (eds) (1993) *Place and the Politics of Identity*. London: Routledge.

Kling, J. and Posner, P. (eds) (1990) *Dilemmas of Activism*. Philadelphia, MS: Temple University Press.

Lees, L. (2003) 'Visions of 'urban renaissance': the Urban Task Force report and the urban White Paper', in R. Imrie and M. Raco (eds), *Urban Renaissance? New Labour, Community and Urban Policy*. Policy Press: Bristol. pp. 61–82.

Levitas, R. (2000) 'Community, utopia and New Labour', *Local Economy*, 15: 188–97.

Little, A. (2002) *The Politics of Community*. Edinburgh: Edinburgh University Press.

Littman, D. and Wills, J. (2002) 'Community of interests', *Red Pepper*, February, pp. 23–5.

Lowe, S. (1986) *Urban Social Movements: The City after Castells*. London: Macmillan.

MacDonald. S. (2002) 'RSLs and tenant groups brought into Elephant & Castle talks', *Housing Today*, 18 July: 7.

Miller, B. (2000) *Geography and Social Movements*. Minneapolis, MN: University of Minnesota Press.

North, P. (2000) 'Is there space for organisation from below within the UK Government's Action Zones?: a test of "collaborative planning"', *Urban Studies*, 37: 1261–78.

Pile, S. and Keith, M. (eds) (1997) *Geographies of Resistance*. London: Routledge.

Putnam, R. (2000) *Bowling Alone: The Collapse and Revival of the American Community*. New York: Simon & Schuster.

Smith, N. (1993) 'Homeless/global: scaling places', in J. Bird, B. Curtis, T. Putnam, G. Robertson and L. Tickner (eds), *Mapping the Futures: Local Cultures, Global Change*. New York: Routledge. pp. 87–119.

Tatchell, P. (1983) *The Battle for Bermondsey*. London: Heretic Books.

Tilly, C. (1974) 'Do communities act?' in M. P. Effrat (ed.), *The Community: Approaches and Applications*. New York: The Free Press.

Wehner, P. (2002) 'Profits wrangle proves to be Elephant's graveyard', *Estates Gazette*, 13 April: 39.

Wilson, E. (1992) *The Sphinx in the City*. Berkeley, CA: University of California Press.

Young, I. M. (1990) *Justice and the Politics of Difference*. Princeton, NJ: Princeton University Press.

Foucault argued that freedom is a practice, a form of embodied action in a particular social and spatial context. In recent years such arguments have been extended into discussions of performative landscapes where everyday people call the landscape into being as they make it relevant for their own lives, strategies, and projects. The following chapters focus on performative responses in/to the city, responses that are or may become emancipatory practices. The authors attentions focus on practices rather than the operation of hidden structures. However, emancipatory practices more often than not involve an in-built critique of hidden structures – such as the capitalist city. Attention focuses on the polities of space in the city, on the counter-production of spaces, on interventions in the city, on the insignificant, the intimate, the mundane, or the eccentric. They remind us that urban space can be ambivalent in terms of social justice and political emancipation.

Gavin Brown explores two sets of (often fleeting) contemporary urban spaces that demonstrate some of the emancipatory potential of the city – sites of public (homo)sex and the carnivalesque spaces created by the direct actions of the anti-capitalist *dis*organization. Reclaim the Streets (RTS). Just as the actions of RTS consciously attempt to undermine, however briefly, the privatization of public space and the alienated compartmentalism of our lives, the liminal spaces of public (homo)sex transcend the private/public binary through which contemporary sexual citizenship is defined. Although the spaces created by RTS actions and public (homo)sex challenge certain societal norms and demonstrate the potential for new forms of communality, Brown questions the extent to which these spaces and the acts that constitute them are sufficient for promoting real and emanciparoty change.

Gary Bridge argues that urban rationality has traditionally been seen as the enemy of the emancipatory city. Rational social relations have been likened to the exchange value of commodities. For Simmel the 'mask of rationality' resulted in a blasé attitude and thus indifference in the city. Administrative of instrumental rationality the spatial and psychological grid imposed on the city has been seen as controlling and repressive. In contrast to these ideas, Bridge asserts strategic rationality – our improvised and performative response to the unpredictable encounters with strangers whilst out walking in the city – as a new way to expand communication and aid in the creation of an emancipatory public realm. Strategic rationality, the rational expectations between the speaker and the audience that enables the speech act to come off, is at its height, he argues, in the spaces of unpredictability in the city. Such an understanding of the dynamics of social interaction is important both theoretically and empirically, not least in the UK where the British Government has recently promoted walking on city streets as a facilitator of social mixing and urban renaissance.

David Pinder focuses on the attitudes and practices of Guy Debord and the Situationist International (SI) as they sought to distance themselves from the normal social encounter in the city. He is especially interested in the ambivalence that they displayed towards the city, for they saw the city as both a site of control and possible sphere of emancipation. He outlines their critique of capitalistic urbanism and their interventionist strategy – new types of games that would expand life and enhance emancipation, for example, the dérive. Space was socially produced and could be socially *re*produced in challenging and critical ways. A review of such modernist and avant-garde movements is important to see what we might learn from the artistic and political practices they used in an effort to open up emancipatory possibilities in the city.

Quentin Stevens uses the concept of play as a framework for considering how we might transgress the social norms encapsulated in city life. Dividing play behaviour into four types – competition, simulation, chance, and vertigo, he examines how each enables escape from the conventions of everyday life. In an account of play as resistance to the spatial and representational regulation of the public spaces of downtown Melbourne, Australia, Steven's pays great attention to the detail of life on the street. He scrutinizes a game of giant chess, a Critical Mass 'bike lift', the public's interaction with public artwork, the Moomba street parade and the Birdman's protest, a man holding up a placard, a dancing man, and in-line skaters. Stevens is positive about such play with the order of things, about the concrete impacts of soft cities upon the hard.

6

Sites of Public (Homo)Sex and the Carnivalesque Spaces of Reclaim the Streets

Gavin Brown

In this chapter I consider two sets of contemporary urban spaces and the spatio-temporal processes through which they are constituted: the carnivalesque spaces created by Reclaim The Streets (RTS) street parties and direct actions, and the (often fleeting) sites of public (homo)sex. While both sets of sites are firmly rooted in the present, they problematize contemporary social relations and hint at new possibilities for future ways of organising (aspects of) society. Although the spaces created by RTS actions and for public (homo)sex challenge certain societal norms and demonstrate the potential for new forms of communality, I question the extent to which these spaces and the acts that constitute them are capable of promoting real and sustainable emancipatory change. It is the tensions and contradictions contained within these sites that I will explore in the pages that follow.

Reclaim the Streets is a radical direct action *dis*organization inspired (partly) by the Situationists (see Pinder, this volume) and born out of the anti-roads protests in Britain in the early 1990s. Although the form of the actions they undertake continues to mutate, they are best known for organizing street parties and road blocks that challenge 'car culture' and the atomized alienation of capitalist society. As I will explore later, the RTS street parties are very much about making serious politics fun and celebrating human creativity and potential in the here and now rather than waiting for a total transformation of society and the relations of production.

I would suggest here that public (homo)sex, as a celebration of queer diversity and sexual pleasures, can fulfil a similar function (albeit less consciously). Just as the actions of RTS consciously attempt to undermine (however briefly) the privatization of public space and the alienated compartmentalism of our lives, the liminal spaces of public (homo)sex transcend the private/public binary through which contemporary sexual citizenship is defined.

The sites discussed in this chapter are created by social, cultural, and political practices. They are festive sites in which carnivalesque processes are at work. They hold no place for dogma and authoritarianism (Bakhtin, 1984), they are times and places of excess and exuberance where (potentially) anything goes. However, they are also firmly rooted in everyday lived experience – they are born out of the alienation and insecurity inherent in capitalist society – and differ from everyday life only in their intensity (Lefebvre, 1991). They simultaneously reinforce social identities formed in the context of that society (albeit oppositional ones) and provide the conditions in which contemporary identities and social positions can become mutated and combined in new ways that contain within them the buds of alternative, less alienating ways of being.

The sites and spaces that I discuss in this chapter can be thought of as examples of Michel Foucault's heterotopia (1986) or as elements of what Hakim Bey (1991), writing from a more avowedly activist perspective, has called 'Temporary Autonomous Zones' (TAZ). Heterotopia are spaces of alternate order that organise elements of the social world differently to that which surrounds them and hint at alternative ways of ordering society (see Lees, 1997). This

enables us to look upon the multiple forms of deviant and transgressive behaviours and politics that occur in urban spaces ... as valid and potentially meaningful reassertions of some kind of right to shape parts of the city in a different image. It forces us to recognize how important it is to have spaces (the jazz club, the dance hall, the communal garden) within which life is experienced differently. (Harvey, 2000: 184)

This, in part, is the rationale for the Reclaim The Streets parties; but their political ethos, which is summarized in the following statement, goes further:

It's about reclaiming the streets as public inclusive space from the private exclusive use of the car. But we believe in this as a broader principle, taking back those things which have been enclosed within capitalist circulation and returning them to collective use as a *commons*. (RTS Agitprop quoted in Jordan, 1998: 140, emphasis added)

And herein lies much of the linkage that I claim between the carnivalesque spaces created by RTS actions and sites of public (homo)sex. In its most literal sense, outdoor homoerotic cruising frequently takes place in parks, on beaches and in publicly accessible woodland – common land. But, beyond that, I argue that these sites foster new forms of homoerotic communality that can potentially contribute to a re-evaluation of meaningful human interaction and community formation through which the reshaping of public space – as commons – can take place.

However, this presents us with a significant tension to be found within the sites and practices under discussion here. As Loretta Lees notes, 'although the publicness of space is legitimated by its unmarked universality, it depends upon particular *constructions* of the public and the proper sphere of its activities

(1997: 322, emphasis in original)'. Just as the notion of 'public space' assumes a false universality that excludes many, so too do these countersites. For, although these sites are created by the capacity of marginalized groups to resist and subvert hegemonic forces, there are still limits to their inclusiveness. The sites of public (homo)sex discussed in this chapter are male-only spaces. The direct actions of RTS have not tended to draw in significant numbers of participants from outside existing urban radical and counter-cultural milieux. These contradictions and ambivalences are addressed in greater detail in the concluding section of this piece.

I begin my discussion with a brief history of Reclaim The Streets and the changing shape of the protests that have been organized and inspired. This is followed by a more detailed description of the carnivalesque spaces created by their street parties and a consideration of the political philosophy that underpins their actions. I conclude my discussion of RTS by examining some of the limitations of their tactical frivolity.

In contrast to the spectacular spaces created by the occasional RTS street parties, I move on to an exploration of the spaces of public (homo)sex, considering these as an example of the carnivalesque brought into the sphere of everyday life. Through a consideration of the site-specific practices that create space for public (homo)sex, I explore how these sites allow queer men to relate in new ways that can destabilize essentialist notions of fixed sexual identities. As with my other case study, I end this discussion by acknowledging the limits to the emancipatory potential of these sites.

Reclaiming the streets

The spaces created by RTS street parties are consciously imbued with a utopian and dissident spirit whose very form is meant as a critique of globalized capitalism and as a signpost to an alternative future.

Reclaim The Streets was originally formed in London in 1991, but quickly became absorbed into the campaign against the extension of the M11 motorway in East London only to be re-formed in 1995 after the defeat of that movement with the final eviction of the Claremont Road encampment. The first RTS street party burst into life one Sunday afternoon in May 1995 when two cars careened into each other on Camden High Street in North London and, as their drivers set about each other's cars with hammers, a crowd of 500 people emerged from among the shoppers to redecorate the street and party the afternoon away to the tunes of a mobile sound system. The next couple of years saw RTS team up with striking tube workers in London and the sacked Liverpool dockers and their families. These links culminated in the March for Social Justice on 12 April 1997 just prior to that year's General Election. Under the banner 'Never Mind the Ballots, Reclaim the Streets', RTS came together with the dockers and striking workers from

Hillingdon Hospital and Magnet kitchens to highlight the need for radical social change that so many hoped the (then) imminent end of two decades of Tory rule would bring about.

Although RTS-inspired street parties continued to occur around Britain and across the globe over the next couple of years, there was no further high profile event in London until June 1999. Then RTS came together with a host of other direct action groups to stage the J18 Carnival Against Capitalism in the City of London. Since J18, RTS have also played a part in organizing other protests, such as the Guerrilla Gardening event in Parliament Square on May Day 2000 (see Pile, this volume). Although the mass media has made much of RTS's involvement in subsequent May Day protests and anti-capitalist mobilizations in Prague, Gothenburg, and elsewhere, these protests were qualitatively different from the 'classic' RTS actions discussed here. Those demonstrations were organized by broader activist coalitions and sought to confront the institutions of global capitalism directly rather than (just) playfully transforming public space.

This brief and somewhat perfunctory history of Reclaim the Streets does little to explain the philosophy and intentions behind the actions or to give even the sketchiest feel for their atmosphere and the spaces they create. It is to this task that I now turn, paying specific attention to the J18 Carnival Against Capitalism and the Guerrilla Gardening event.

A carnival needs masks

It is difficult in a few sentences to capture the vitality, creativity, and festive fun of a Reclaim The Streets event. Nevertheless, I want to provide readers who have not been part of one with a flavour of them. I acknowledge that my observations here are very much based on the London actions. The street parties that have taken place in cities as diverse as Bristol and Hull in England, or further afield in Berlin, Prague, or San Francisco, will have almost certainly produced different results reflecting the various cultures, political climates, and urban landscapes in which they happened.

The symbols and instruments of carnivals and festivals across the world are a prominent feature of the London street parties: from the maypole erected on the tarmac outside Parliament at the Guerrilla Gardening event to the omnipresent samba band whose persistent rhythms helped gel and direct the crowd. As in most carnivals, costumes and masks have come to play an important part in the parties. The role of the mask was most obvious in the J18 Carnival Against Capitalism, where red, green, black, and gold masks were distributed among the crowd as it assembled on the forecourt of Liverpool Street Station. In part, the masks were meant to serve as a means of easily dividing up the crowd into four colour-coded contingents that could be moved along different routes by members of those affinity groups who were 'in the know' to confuse the police's efforts to prevent the event taking place and thus maximise the chances of reaching the planned venue for the day's main party (i.e. the LIFFE futures exchange building). Printed on the reverse of the masks was an explanation of their significance that served as a statement of intent for the day and the spirit of Reclaim The Streets:

Those in authority fear the mask for their power partly resides in identifying, stamping and cataloguing: in knowing who *you* are. But a Carnival needs masks, thousands of masks; and our masks are not to conceal our identity but to reveal it ...

The masquerade has always been an essential part of Carnival. Dressing up and disguise, the blurring of identities and boundaries, transformation, transgression; all are brought together in the wearing of masks. Masking up releases our commonality, enables us to act together, to shout as one to those who rule and divide us 'we are all fools, deviants, outcasts, clowns and criminals'. ...

While the gangs of state and capital become evermore faceless their fear of the faces of everyday resistance grows ... Today we shall give this resistance a face; for by putting on our masks we reveal our unity.

At the Guerrilla Gardening event, and very much in the spirit of the transgressive 'blurring of identities and boundaries' proclaimed by the text on the J18 masks, was a man wearing a horned leather bondage mask adorned with a crown of ivy leaves. His costume evoking thoughts of the 'Green Man' of pagan myth and serving as a reminder that before becoming International Workers' Day, the May Day festivities had their roots in ancient fertility rituals. As one of the banners hung above the square that day proclaimed, 'Resistance is Fertile'.

Finally, the parties not only take control of the street for their duration, they alter that space in creative and symbolic ways. At one early street party the road was transformed into a beach, complete with sand and deck chairs. On J18 a fire hydrant was opened sending a fountain of water several storeys high into the air and creating an impromptu paddling pool in which the revellers could gain some relief from the hot summer sun. The Guerrilla Gardening event was clearly organized with the intention of demonstrating how dead, inhuman, and unproductive much of what passes as public space in our cities is; and, with Parliament Square's turf re-placed on the tarmac and vegetation planted in its place, the point was made.

Celebrating dissent and dissenting celebrations

In two distinct, but ultimately inseparable ways, the RTS street parties distinguish themselves from the dour seriousness of the orthodox Left. First, through their heady mix of music, theatricality and art, the street parties and other actions raise serious and pressing political concerns in an atmosphere of exuberant fun. Second, they are more than just protests *against* something, they are celebrations of dissent and the potential of human imagination that does not defer social change to some future date, but attempts to enact it in the here and now. Their's is not so much a goal-oriented politics, but one in which a process of conscious self-activity is seen as centrally important and the seed of future social revolution.

It is through this philosophy that RTS harnesses the true spirit of the carnival for their actions. A carnival represents and celebrates a temporary liberation from the established order in which social hierarchy and norms are suspended. At the carnival, crowds of people take control of the streets to celebrate their own creativity and desires. Most importantly, carnivals are participatory events – it is no fun, and often not possible, just to stand on the sidelines and watch. The crowd has a habit of engulfing you in its exuberance. There is no separation between audience and performers. It is created by and for everyone. The linkages between the traditional carnival and the political philosophy underpinning Reclaim The Streets should be clear:

> [T]he Street Party can be read as ... an attempt to make Carnival *the* revolutionary moment. Placing 'what could be' in the path of 'what is' and celebrating the 'here and now' in the road of the rush for 'there and later', it hopes to re-energise the possibility of radical change ... It is an expansive desire; for freedom, for creativity; to truly live. (Bailie, 1997: 5, quoted in Jordan, 1998: 141)

To allow people to assemble freely in the streets is always to flirt with improvisation and the risk of the new and unexpected. This is as true of the ways in which queer men undermine heteronormative moral codes by cruising public spaces as it is of the possibility that revelry might unleash rioting. In ancient societies the concept of the jubilee and the saturnalia originated in a belief that some events should exist outside the scope of regulated 'profane' time, quite literally in gaps in the official calendar (Bey, 1991: 105).

In a world where socialising takes place increasingly in privately-owned spaces such as shopping malls, coffee shops, and bars rather than on the porch or in the city square, actions such as the RTS street parties offer us an opportunity to re-evaluate and actively change the ways in which we relate to public space and each other. But furthermore, as the rate of technological advance and the cycles of capital accumulation force us to live our lives at an ever faster pace, I would suggest that the RTS street parties also provoke new ways of experiencing time, not just space. John Jordan (1998: 133) has written that 'the inherent risk, excitement and danger of the action creates a magically focused moment, a peak experience, where real time suddenly stands still and a certain shift in consciousness can occur.' As I discuss later in this chapter, a similar sense of 'timelessness' is experienced by many men while cruising.

Limitations: what if the street is a cul-de-sac?

So far, my commentary on Reclaim The Streets has been largely positive, but there are some very real problems with their actions and the way they organize. As George McKay (1998: 50) has noted, RTS have as yet not taken any action that challenges or addresses the gendered experience of (fear in) public space. Similarly, most of the street parties have remained largely white affairs and this has allowed the police to target black activists for arrest and prosecution – the ethnic mix of the revellers' mugshots posted on the City of London police's website following the Carnival Against Capitalism did not seem to reflect the overwhelming whiteness of the crowd on the day.

One issue that I have so far largely ignored is the question of violence. The street parties are intended to be non-violent affairs and yet the three major events in London since J18 have all ended violently. At times sections of the crowd have indulged in (what the mainstream media has characterized as) 'random' acts of destruction against symbols of corporate power that have provided the police with a justification for the violent dispersal of the gathering. To my mind, RTS's methods of organisation do little to protect the crowd in such circumstances. They make a virtue of being a *dis*organization with no central leadership in which loosely linked autonomous affinity groups take responsibility for aspects of the preparations for an event and their own actions on the day. There is a political logic to this organizational approach: it fits neatly in with their commitment to self-activity and a non-hierarchical means of organising society. The approach also has some practical advantages, especially in terms of security, for the type of actions that RTS organise, but it is insufficient for neutralizing voluntarist elements in the crowd or providing organised defence of the protests when faced with all-out assaults by riot police.[1]

To date, the response of RTS to these pressures has been to keep on their toes and to move into new forms of organising. These activists seem to have taken to heart Bey's theory of the Temporary Autonomous Zone (TAZ).

> The TAZ is like an uprising which does not engage directly with the State, a guerrilla operation which liberates an area (of land, of time, of imagination) and then dissolves itself to re-form elsewhere/elsewhen *before* the State can crush it. ... In sum, realism demands not only that we give up *waiting* for 'the Revolution' but also that we give up *wanting* it. 'Uprising,' yes – as often as possible and even at the risk of violence. The *spasming* of the Simulated State will be 'spectacular,' but in most cases the best and most radical tactic will be to refuse to engage in spectacular violence, to *withdraw* from the act of simulation, to disappear. (Bey, 1991: 101)

Sadly the corollary of 'placing "what could be" in the path of "what is" and celebrating the "here and now" in the road of the rush for "there and later"' (Bailie, 1997: 5) seems to be abdicating all responsibility for helping to bring about the 'there and later'. And therein lies a major problem with the RTS-style street parties: they provide a spark and an inspiration for alternative modes of living, but offer no serious clues as to how to achieve them.

In part this is a result of the anarchist inspired politics of RTS activists, but it also results from a misunderstanding of Bakhtin's concept of carnival. As Julian Holloway and James Kneale (2000) have recently stressed Carnival is not an abstract 'force' but a set of practices located in specific contexts with indeterminate consequences. It is important to move away from seeing Carnival as an *inversion* of order to prevent an episodic view of cultural politics, where disorder and transgression are restricted to rare, large-scale outbursts of popular feeling.

> Bakhtin made it plain that carnival was not simply to be found in revelry or riots, but also in everyday speech, conceptions of the body, and so on. As the dialogical Other of official culture, Carnival must always be present; it contaminates the supposedly monological utterances of the powerful. Carnival may be a weakened force, but its currents still run through popular culture. In this sense, we should be looking for elements of everyday life which can become 'carnivalised'. (Holloway and Kneale, 2000: 81)

The large-scale revelry (and riots) unleashed by RTS street parties are rare events, but every hour of every day queer men engage in public (homo)sex in the saunas, public toilets, and parks of even the smallest city. As such, public (homo)sex may represent an aspect of everyday life 'carnivalised', and I turn now to a consideration of the emancipatory potential of those sites and practices.

Sites of public (homo)sex

In this section I consider contemporary public sex environments (PSEs), such as parks, common land, cemeteries and public toilets ('cottages' or 'tearooms' depending on which side of the Atlantic you are) that gay men appropriate for erotic pleasure, and commercial public sex venues (PSVs), such as saunas, sex clubs and backroom bars specifically created to cater to those desires. After elaborating on the differences in form and function between these two sets of spaces, I consider how they operate as heterotopic sites that create space for multiple forms of deviant and transgressive behaviour and allow queer men the opportunity to experience life differently.

Erotic topographies

Public sex environments are created (however temporarily) by a social process, namely cruising, that subverts the design and intended use of existing sites. In these sites cruising and (homo)sex takes place incidentally alongside the more 'legitimate' uses for which the sites were designed. In contrast, PSVs are spaces designed to facilitate the replication of (and, at times, expand upon) the repertoire of acts that take place in PSEs. The owners of these commercial venues actively promote (or at least turn a blind eye to) sex on their premises.

John Hollister (1999) has suggested that the term 'public sex' is a misnomer. On one level he is correct, public (homo)sex seldom takes place in open public view. Cruisers using PSEs usually try to ensure that any sex play takes place out of the public view under cover of darkness or foliage. Only the most foolhardy, drunken, or exhibitionist cruiser will risk engaging openly in a sex act in front of someone he is not convinced is also cruising. Still, these sites are public in the sense that potentially anyone can access them. The publicness of public sex venues is less straightforward. Entrance to these sites is usually controlled in some way, most frequently through a nominal membership system or an entrance charge, and there is almost no risk of an unsuspecting member of the public stumbling into them by mistake. However, in legal terms (at least according to current English law) these are public spaces and the fact that any sex acts that may take place within them may involve or be viewed by more than two consenting adults marks them as transgressive sites on the fringes of legality. It is for this reason, as well as the broad similarity of the practices pursued in these sites, that I consider PSVs here alongside more strictly *public* sex environments.

The spatial distribution of both PSEs and (to an even greater extent) PSVs, as well as the ability of men to access them, is uneven. The privatization and curtailment of public space has led to the closure and/or erasure of many public sex environments in British cities. Frequently one of the first services downsizing local authorities have cut is the provision of public toilets. At a stroke this has re-drawn the erotic topography of many urban areas. At the same time British cities have increasingly seen the commodification of public (homo)sex through the growth of commercial sex venues. After a period of intense police harassment at the height of the AIDS moral panic in the late 1980s, such venues are now largely left to their own devices, so long as they 'play the game' and comply with fire regulations and other health and safety concerns. This has enabled club proprietors to invest more confidently in their premises and public (homo)sex is now a profitable business.

By placing homoerotic activity squarely in the public sphere, cruising grounds, saunas, and sex clubs challenge the patriarchal and heteronormative assertion that sexual activity (especially that of sexual dissidents) should be reserved for the 'privacy' of the home.

The choreography of cruising

In exploring the emancipatory potential of public (homo)sex, it should be remembered that cruising sites vary too much by location and time of day to make easy generalisations about them. As Hollister (1999: 58) comments, 'cruising is not just a repertoire of techniques. It cannot be separated from the locations where it takes place.' Cruising may take place in just about any location, but men will have a far greater chance of success in locations where other men expect to find companionship and sex. In such sites men communicate with each other and consummate their (brief) affairs using the props offered to them by the sites in which they find themselves. In this way refiguring space is central to all cruising. Cruisers assess other men's sexual availability on the basis of how they approach and occupy the cruising site, just as they themselves use those spaces in ways that signal their own availability to others. This ritualized repertoire of gestures and glances is superficially simple, but in the context of the cottage or steamroom intricate enough to convey meanings as complex as sexual desire, preferred sex role, and consent (Hollister, 1999: 60).

Just as the way in which a man moves and holds his body plays an important function in the choreography of cruising, so too can the manner in which he is dressed. Men use 'costume' as a means of presenting their desires and manipulating those of others. This is most obvious in the costumes worn in fetish clubs, but can be just as important in a cottage or an outdoor cruising area – sportswear, workboots or a crisp suit can all contribute to whether and how a scene will develop and the roles that respective men will take in that scene. Like the carnival masks discussed earlier in this chapter, in the context of homoerotic cruising dressing up can lead to the blurring of identities and boundaries. Men use the props of everyday life to transform and transgress the ordinary.

I have already noted that only the most foolhardy cruiser will risk engaging openly in a sex act in front of someone he has not already assessed as another cruiser. In its own way this, like the use of costume and 'masks', highlights the carnivalesque in public (homo)sex. Carnival is a participatory event because 'its very idea embraces all of the people. While Carnival lasts, there is no other life outside it' (Bakhtin, 1984: 7). In a PSE a non-participant is potentially a policeman or a queer-basher and can squash any action taking place, while the presence of a voyeur will often change the dynamic of play.

Only in the most secure sites will participants risk communicating their desires verbally, but even in gay saunas overt conversation is frequently reserved for the television lounge rather than those spaces set aside for cruising. While the 'silence' of the cottage or the orgy room might be seen by some as a mask of shame, it allows men to be generous in evaluating their prospective partners as desirable, especially when they can barely see each other.

Outdoor cruising areas in parks, woodland, and beaches allow men to indulge in and explore a broader range of roles than they might in a small toilet that can only accommodate a handful of men at any one time. However, in saunas and other commercial sites participants do not need to calculate who is there to cruise and who should be shielded from any cruising that is going on before making an advance. They need only establish who is available and interested. On those occasions (most frequently, late at night) when everybody at a cruising site is there to cruise, the commons come to resemble saunas more closely with the same social etiquette of ushering in men that one finds attractive and warding off those that one does not.

Homoerotic communalities

Given the tendency for sexual desire and availability to be communicated by non-verbal means, how a man understands or presents his sexual identity in other spheres of his life tends to be of secondary importance to how he acts in the cruising zone. Far more crucial than whether he identifies himself as gay, straight, or bisexual is his ability to use a given cruising site effectively and the extent to which he is perceived to be desirable by any other men present at the time (Hollister, 1999: 63). It is in this tendency to disrupt the homo/heterosexual binary and fixed social identities constructed around it that the emancipatory potential of such sites can be found. The practice of public (homo)sex is a creative act through which new spaces for sex and new ways of being-in-the-world are produced. As Ira Tattelman has described in his ethnography of a New York gay bathhouse:

> The baths provide a public place where a wide mix of strangers can come together. Men from vastly different emotional, sexual and physical worlds arrive at the baths wanting to make connections with other men. Tolerant of difference, open to diversity of uses, the public territory of the bathhouse gives men the space to define, support or flaunt their sexual interests. (1999: 71)

If the baths and other sites of public (homo)sex problematize essentialist notions of fixed sexual identities through their concentration on acts rather than claimed identities, then we must also remember that they have played an important role in the formation of gay identities (and continue to do so). At a time when most gay rights activists are pursuing an 'equal rights' agenda that unquestioningly claims, for lesbian and gay people, the same rights and privileges as our (middle class) heterosexual peers, it is easy to forget that the politics of early gay liberationists were very different. In the 1970s, early gay liberation praxis was more concerned with challenging the centrality of the nuclear family and liberating all human sexuality than achieving equality within bourgeois society. Early gay liberation praxis placed an emphasis on sexual expression as a signifier of a liberated existence and, in this context, the emotional and physical security of the baths allowed men to celebrate new social structures and pleasures through communal sex. For many, the act of making a sexual choice in front of others, who by their presence were celebrating and endorsing those choices, fostered a new spirit of self-sufficiency and a redefinition of what it meant to be gay. It could, therefore, be argued that spaces of public (homo)sex played a strategic role in the development of modern gay identities as they not only confronted heteronormative social mores, but promoted sexual diversity and proposed alternative forms of communality between men (Tattelman, 1999).

Public (homo)sex simultaneously blurs and reinforces the hetero/homosexual binary. These practices privilege (homoerotic) behaviour over queer identities, but at the same time help to construct and reinforce those identities. Although a man who does not think of himself as, at least a little, queer is unlikely to visit a commercial queer public sex venue, horny 'straight' men have a habit of getting caught up in the excitement of PSEs and this can destabilize previously assumed identities. In a homophobic society this can, of course, be deeply unsettling for some men, but for others it can be a liberating experience. But even for men who have already claimed a queer identity, the experience of public (homo)sex, especially when it takes place in a communal context, can open up new ways of giving and experiencing love and affection that are not based on existing heteronormative models. As Scott O'Hara explains:

> On those occasions (long ago and far away, now, alas) when I was bent over the bench in the orgy room at Manscounty, being fucked by one man after another, I do think there was more love involved than most people would be willing to admit. Love is very intimately involved with the giving of pleasure; sex is simply one of the easiest ways to give pleasure to strangers. (1997: 73)

Although public sex environments have an etiquette of their own, within these spaces the conventional rules and regulations of sexual 'normality' are removed. Such sites are a playground offering men a place in which to relax and indulge their homoerotic desires away from the realities of a homophobic society. They also offer a respite from the 'mundane niceties of home-type sex' (O'Hara, 1997: 139): are my sheets clean? What will I make him for breakfast? Scott O'Hara further linked this sense of removal from the petty worries of everyday life to the architecture and interior design of gay saunas. For him the almost universal lack of windows in them and the absence of any clocks from the 'action zones' reinforced a sense of separation from the outside world. Although this exaggerated distance from mainstream society is consciously articulated in the design of many commercial public sex venues, many men experience a similar sense of dislocation in 'outdoor' cruising sites, where time seems to stand still, or at least pass at a slower pace.[2] As was noted in the earlier discussion of the RTS street parties, this is a common experience in heterotopic spaces. Nonetheless, there are limits to the emancipatory potential of these sites and it is to these limitations that I now turn.

Erotic alienations

Tattelman (1999: 77) has argued that the spatial exclusivity of the cruising site encourages the interaction of 'otherwise unrelated bodies' in ways that begin to undermine the feelings of isolation and alienation fostered by hetero-normative capitalist society. However, because mainstream society teaches men that the act of being seduced represents a loss of control and that self-control is everything, it would be naïve to believe that issues of stigma, alienation, and personal shame are never reproduced within cruising sites. Cruisers do not just have to contend with the pressures of institutional homophobia and societal disapproval, but with the prejudices of other site users:

> In these communal spaces, the nature of desire can also become rude and demeaning. Young men can be ruthless in their rejection of older men. At the same time, those less attractive can persistently grab at the more attractive to the point of humiliation. Issues of race and class are often reproduced inside the baths and exaggerated or campy mannerisms are discouraged. (Tattelman, 1999: 87)

Despite London's rich cultural diversity, much of the commercial gay scene in this 'global' city remains distinctly 'white'. Cottages and open air cruising zones are a different matter. In these sites language barriers and economic disadvantage are less of an issue; and paradoxically, given the more public nature of these spaces, the stigma of (potential) exposure is lessened as one's interest in making a homoerotic connection can be masked by more 'legitimate' uses of the site.

It is also important to note the gendered reality of these spaces. Many of the sites appropriated for public (homo)sex, public toilets as well as many of the commercial venues under discussion here, are designated as male-only spaces. In those more openly accessible cruising sites such as parks and common land, the presence of a woman is likely to interrupt any sexual activity taking place. It is worth remembering also that as cruising often takes place most openly in these outdoor sites after dark, the likelihood of women stumbling across the action is much reduced. Women's sexual conduct is still more heavily regulated than men's, as is their use of public space, and this partly explains why there are so few equivalent spaces in which (queer) women can enjoy communal homoerotic experiences.[3]

Clearly the emancipatory potential of these sites, like their spatial distribution and the ability of men to access them, is uneven. These inequities need to be addressed further by queer geographers. The mix of performative activities available in any given place and time are not independent of the physical, social, and economic environment in which they take place (Harvey, 2000: 98). While (some) sites of public (homo)sex may offer up possibilities for future forms of communality, they are still very much rooted in the present and limited by the realities of contemporary social relations. Public (homo)sex may offer us episodes of release and hedonistic enjoyment, through which new modes of living can occasionally be glimpsed, but they do little to confront and challenge the homophobia of heteronormative society. Few men, if any, are likely to construct a sexual identity purely on the basis of public sex encounters. Even if this were the case, any resulting identities would not be formed in a social vacuum: the gay identities that were forged and consolidated during the 1970s in the baths of New York and other cities have mutated in the face of AIDS and the shifting political climate of the intervening quarter of a century. Time spent in the baths or on a cruising pitch may offer glimpses of new and less alienated forms of human interaction but, as Tattelman (1999: 91) so poignantly reminds us, '[w]hen it is time to get dressed, one feels constrained; one's "street" identity returns all too quickly.'

Strategic sites and spaces of hope

My discussion of the spaces of public (homo)sex and radical direct action has been largely a celebratory and optimistic one. Both sets of spaces take 'the alienated, lonely body of technocratic culture and transform it into a connected, communicative body embedded in society' (Jordan, 1998: 134). They do so through processes of creative resistance that rely on intuition and imagination more than reason and rationality. By (partially) transcending societal norms these spaces and the social processes that create them assist in the construction of new collective identities (while destabilizing existing ones) and of communities of action through which both the personal and the political are translated onto a broader terrain of human action. As these alternative collectivities are created so too are new spaces that shape 'the political person as well as the ways in which the personal is and can be political' (Harvey, 2000: 241).

In contrast to this optimistic, utopian view though, it is possible to articulate a more pessimistic perspective. As Harvey (2000: 237) has highlighted in his recent meditations on the future of utopian praxis, the very pace of contemporary life, in which just about every aspect of human existence is subsumed by the cycles of capital accumulation, 'preclude[s] time to imagine or construct alternatives other than those forced unthinkingly upon us as we rush to perform our respective professional roles'. In these grim circumstances many 'alternative lifestyles' or new modes of self-expression either become recuperated into the process of capital accumulation or repressed to one degree or another. Indeed, police repression, low-level harassment, and press sensationalism have forced direct actionists to keep on their toes and constantly find new modes of operation. As a result the Reclaim The Streets street parties discussed in this chapter may already be a feature of history.

The growth of public sex venues in many major British cities over the last decade points to the (partial) recuperation of public (homo)sex. Although what was once an uncommodified public experience is increasingly been harnessed for private profit, I would argue that gay saunas and sex clubs are still heterotopic spaces as they facilitate transgressive behaviour and allow queer men to experience, for a few hours, life made extraordinary. This highlights an important, but seldom recognized, point that is at the heart of debates about the (post)modern city – it is entirely possible for utopian processes of creative resistance to survive and coexist in the same spaces as more dystopic processes of capital accumulation and repression (Lees, 1997). The time is perhaps long overdue to move beyond an either/or view of the city as emancipatory/revanchist and begin to develop strategies that can offer ways out of this antinomy.

The revanchist city thesis (Smith, 1996) tends to overestimate the impact of the very material processes of repression, reification, and capital accumulation that are at work in contemporary society, while underplaying the capacity of human beings for individual and collective acts of resistance to these alienating social and economic processes. The carnivalesque spaces that I have examined in this chapter resist dogma, authoritarianism, and narrow-minded seriousness. They are 'opposed to all that is finished and polished, to all pomposity, to every ready-made solution in the sphere of thought and world outlook' (Bakhtin, 1984: 3). As such they are prime examples of our capacity for creativity and resistance in the face of homogeneity and repression. However, fun as these spaces are, it is not enough simply to celebrate the endless, open process of transgression presented by the carnivalesque. We must find ways of harnessing that spirit of exuberant celebration and adventure in strategic ways that can change rather than just subvert society.

The spaces of alterity and 'otherness' discussed above are strategic because they allow alternative ways of living to be explored through existing social process, rather than purely as utopian imaginings. Within them alternative modes of communality can begin to take shape and, because they are actually existing sites, they offer the opportunity for a more effective critique of existing social norms and processes. They inspire utopian possibilities that are firmly rooted in the contradictions of existing social relations and identify agencies and processes of change by which these can potentially be realised. Furthermore, they serve to remind us that emancipation must mean more than just economic and social justice for the majority of humanity – that everyday life must be reclaimed for itself and *disalienated*. They suggest it may be possible to achieve 'an *intensification* of everyday life, or as the Surrealists might have said, life's penetration by the marvellous' (Bey, 1991: 111, emphasis in original). They hint at how we might realize Lefebvre's desire for cities that 'provide the means for "free associative" expression' as 'arenas of *jouissance*, of intense sensual and sexual pleasure and excitement' (Merrifield, 2000: 179).

Because these possibilities are exposed through processes that are site-specific, the utopianism that is produced is close to the 'dialectical utopianism' that Harvey (2000) called for in *Spaces of Hope*. This is a utopianism that is 'rooted in our present possibilities at the same time as it points towards different trajectories for human uneven geographical developments' (Harvey, 2000: 196). Of course, there are dangers that the possibilities to be found both in the sites of public (homo)sex and the carnivalesque actions of Reclaim The Streets will get bogged down in endless cycles of transgression and open projects (such as Bey's model of the TAZ) that never come to a point of closure within space and place (Harvey, 2000: 174). Open experimentation with the possibilities of new spatial forms has its merits: it can permit a host of explorations of a wide range of human potentialities as part of an emancipatory strategy. But, if the project remains too open from the start, then it risks total recuperation by the forces of capital accumulation before it can be made a reality. To achieve this level of closure requires collective and individual rituals of resistance that should be both serious and playful in order not to lose hold of the carnivalesque spirit that makes these spaces both meaningful and a joy to experience.

Notes

1 Paradoxically, at the protests against the IMF meeting in Prague in September 2000 and against the G8 summit in Genoa in July 2001, the tactical frivolity of the 'pink bloc' (which is in part inspired by RTS's style of direct action) came closest to breaching the security zones surrounding the summit venues.
2 Thanks to Loretta Lees for pointing out that a similar constructed distance from the 'outside world' exists in many shopping malls.
3 This is not to imply that such spaces do not exist, but to acknowledge that they are fewer and farther between and that women-only public sex venues are seldom promoted as openly as the equivalent spaces used by queer men.

References

Bailie, D. (1997) 'Reclaim the streets' *Do or Die*, 6: 1–10. (The paper is also available online at http://www.eco-action.org/dod/no6/rts.html)

Bakhtin, M. (1984) *Rabelais and His World*, trans. H. Islowsky. Bloomington, IN: Indiana University Press.

Bey, H. (1991) *T. A. Z.: The Temporary Autonomous Zone, Ontological Anarchy, Poetic Terrorism*. New York: Autonomedia.

Foucault, M. (1986) 'Of other spaces', *Diacritics*, 16: 22–7.

Harvey, D. (2000) *Spaces of Hope*. Edinburgh: Edinburgh University Press.

Hollister, J. (1999) 'A highway rest area as a socially-reproducible site', in W. L. Leap (ed.), *Public Sex/Gay Space*. New York: Columbia University Press. pp. 55–70.

Holloway, J. and Kneale, J. (2000) 'Mikhail Bakhtin: dialogics of space', in M. Crang and N. Thrift (eds), *Thinking Space*. London: Routledge. pp. 71–88.

Jordan, J. (1998) 'The art of necessity: the subversive imagination of anti-road protest and Reclaim the Streets', in G. McKay (ed.), *DiY Culture: Party and Protest in Nineties Britain*. London: Verso. pp. 129–51.

Lees, L. (1997) 'Ageographia, heterotopia, and Vancouver's new public library', *Environment & Planning D: Society & Space*, 15: 321–47.

Lefebvre, H. (1991) *Critique of Everyday Life*, trans. J Moore. London: Verso.

McKay, G. (1998) 'DiY culture: notes towards an introduction', in G. McKay (ed.), *DiY Culture: Party and Protest in Nineties Britain*, London: Verso. pp. 1–53.

Merrifield, A. (2000) 'Henri Lefebvre: a socialist in space', in M. Crang and N. Thrift (eds), *Thinking Space*. London: Routledge. pp. 167–82.

O'Hara, S. (1997) *Autopornography: A Memoir of Life in the Lust Lane*. Binghamton, NY: Harrington Park Press.

Smith, N. (1996) *The New Urban Frontier: Gentrification and the Revanchist City*. New York: Routledge.

Tattelman, I. (1999) 'Speaking to the gay bathhouse: communicating in sexually charged spaces', in W. L. Leap (ed.), *Public Sex/Gay Space*. New York: Columbia University Press. pp. 71–94.

7 Inventing New Games: unitary urbanism and the politics of space

David Pinder

[W]hile the history of cities is certainly a history of freedom, it is also a history of tyranny, of State administration controlling not only the country but also the city itself. The towns may have supplied the historical background for the struggle for freedom, but up to now they have not taken possession of that freedom. (Guy Debord, 1994 [1967]: thesis 176)

We must define new desires appropriate to today's possibilities … . We must now undertake an organized collective work that aims for the unitary use of all means of revolutionizing everyday life. That is to say, we must first recognize the interdependence of those means from the perspective of a greater domination of nature, a greater freedom. We have to construct new ambiances that will be simultaneously the product and the instruments of new behaviours. To do this, we need to employ empirically at the outset the everyday processes and cultural forms that now exist, while contesting their inherent value … . Our action on behaviour, in connection with other desirable aspects of a revolution in mores, can be briefly defined as the invention of games of an essentially new type. (Guy Debord, 1985 [1957]: 615, 617)

How to realize the freedoms of cities? How to envisage and construct urban spaces as part of a strategy for social transformation, so that they become both the products and the instruments of emancipation? These questions have long occupied artists, activists, and critical thinkers, among others. Visions of cities of freedom have been prominent in the struggles of modernist and avant-garde groups as well as those on the radical left. In their commitment to social change, such groups have typically taken an antagonistic stance towards existing cities. Yet at the same time, alternative urban visions have often acted as beacons for exploring the emancipatory possibilities of life in the present and breaking through to new and better futures.

Among such groups have been Guy Debord and his colleagues in the Situationist International (SI) who confronted changing forms of social life and urbanisation in the second half of the twentieth-century. Debord (1981a [1956]: 51) once referred to contemporary cities as 'those centres of possibilities and meanings'. Seeking to transform everyday life, the situationists explored such possibilities and charted how they might be opened up. But at the same time they understood how these possibilities were controlled under current conditions, how they were channelled, circumscribed, or even denied by powerful social and political forces whose dominance was dependent upon such suppression. An ambivalent relationship to the city was thus characteristic of situationist projects as it had been for many modernist and avant-garde movements before them. For the situationists the city was simultaneously a key site of control and alienation, through which prevailing capitalist socio-spatial relations of domination are cemented and reproduced, and a realm of possible emancipation, human fulfilment, and play in which people could challenge alienation by creating spaces to fit their own needs and desires and thereby realize their potentialities as living subjects.

In this chapter I want to explore tensions and ambiguities that characterize situationist attitudes to the city as a site of possibility and a potentially emancipatory space. The two quotations from Debord at the head of the chapter provide the frame for my discussion. In the next section I address the situationists' critique of the city as a space of control and the concretization of unequal and repressive social relations. In particular, I focus on the association in much situationist writing of urbanism with fixity and a 'freezing' of life. This critique was informed by Marxist theory and developed through dialogue with Henri Lefebvre, who had a close association with the group in its early years. But in subsequent sections I turn to how the situationists sought to challenge and transform dominant forms of urbanism and their frozen qualities through the 'invention of games of an essentially new type'. In the situationists' view, the struggle for freedom was a revolutionary endeavour that ultimately required the overturning of repressive social and spatial relations, and the production of a space and way of living that fulfils the potentialities that currently lie within the everyday. 'We want the most liberating change of the society and life in which we find ourselves confined,' argued Debord in 1957 at the foundation of the SI. 'We know that such a change is possible through appropriate actions' (1985: 609; tr. 1981b: 17).

A central action for the SI during its early period in the late 1950s and the beginning of the 1960s was 'unitary urbanism', which they presented as part of a political struggle for emancipation. It was based on a crucial political and theoretical insight – since so powerfully emphasized by radical geographers and thinkers such as Lefebvre, David Harvey, Doreen Massey, Edward Soja and yet still in need of wider emphasis and development – that to change everyday life it is also necessary to change space. In this context it is valuable to return to the radical approaches to cities such as those developed by the situationists. This is not simply to show how these approaches anticipate more recent theoretical concerns, despite the lack of attention they have received until relatively recently. More significantly, it is because they continue to pose important political questions about the challenge of imagining and creating urban spaces differently. In focusing on the situationist activity of unitary urbanism, I therefore want to shed light not only on the group's approach to urban questions but also on fundamental issues about the role of emancipatory visions of urbanism in struggles to change life. This is part of a wider study of twentieth century utopian visions of the city in which I am seeking to explore the contested nature of modernist urbanism and to reclaim the significance and potentially inspiring qualities of past forms of opposi-tional and transgressive utopianism. Such a rehabilitation of utopianism is all the more pressing at the present time, in the face of widespread scepticism if not hostility towards the very idea of utopia (see Pinder, 2002, 2005).

Geology of lies

The first quotation at the beginning of this chapter comes from Debord's book *La société du spectacle*, originally published in 1967. It appears in the midst of his critique of urbanism as the appropriation of the environment by capitalism and as the refashioning of space according to dominant interests. Urbanism is here associated not with social connections, encounter, still less with 'games'. Instead, Debord presents its essential reality as isolation. It is based on the suppression of the street, as the culmination of a dream long harboured by administrators and urbanists concerned with instituting their own understanding of urban order and the development of urbanism as 'the technology of *separation itself*' (1994: thesis, emphasis in the original). Debord (1994: thesis) notes: 'Marx considered that one of the bourgeoisie's great merits as a revolutionary class was the fact that it "subjected the country to the rule of the towns" – whose very air made one free.' Yet he insists that this history of urban freedom needs to be understood at the same time as a history of tyranny and control, and that cities have as yet still not truly taken possession of that freedom. Debord indeed laments the curtailment of urban freedoms through the contemporary urge to destroy cities, an urge that he connects to the development of the economy that has led not to the tran-scendence of the split between the town and country, as looked forward to by Marx, but rather to their collapse. Elements of both are now strewn through capitalist societies as urbanism produces 'a *pseudo-countryside*', complete with lifeless new towns, and 'a new, artificial peasantry' that shares its pre-decessor's inability through social and geographical circumstance to become a creative historical force. Debord asserts: 'it is precisely because the libera-tion of history, which must take place in the cities, has not yet occurred, that the forces of *historical absence* have set about designing their own exclusive landscape there' (1994: thesis 177, emphases in the original).

Here Debord is addressing urbanism in terms of 'the spectacle', a term that he and the situationists use to critique the conditions of the post-war period in which the commodity has completed its colonization of social life. The basis of the society of the spectacle, so they argue, lies in the principles of non-intervention and contemplation where life is increasingly alienated and lived through the commodity and the image. In a 1959 film by Debord (1992: 32), a male voice announces what will become a key concern in these later writings: 'That which was directly lived reappears frozen in the distance, fitted into the tastes and illusions of an era carried away with it.' The line is echoed in the second sentence of Debord's book on the spectacle, following an opening line that itself plays on the beginning of Karl Marx's *Capital*:

> The whole of life in those societies in which modern conditions of production prevail presents itself as an immense accumulation of *spectacles*. All that once was directly lived has become mere representation. (1994: thesis 1, emphasis in the original)

The theme of distance and freezing is further taken up in subsequent parts of the book as well as many other situationist writings. According to Debord (1994: thesis 35), the spectacle 'arrogates to itself everything that in human activity exists in a fluid state so as to possess it in a congealed form'. With this assertion, Debord is not denying the significance of change and mutation but suggesting that the endless production of the ever-new upon which capitalism depends masks the solidification of the ever-same in the shifting guises and appearances of the spectacle-commodity form. He states: 'If henceforward the *free space of commodities* is subject at every moment to modification and reconstruction, this is so that it may become ever more identical to itself, and achieve as nearly as possible a perfectly static monotony' (1994: thesis 166, emphasis in the original). Under such conditions, every effort is made through the production of spectacles and pseudo-histories to maintain what Debord (1994: thesis 200) calls 'the equilibrium of the frozen time that presently holds sway.'

In Debord's account urbanism is presented as a key force for such freezing and separation. It atomizes people who have been brought together by the means of urban production so that, instead of forming collective movements, they remain isolated together. While the spectacle unites separate fragments and integrates the population, it unites and integrates them only in their separateness. Debord therefore contrasts urbanism with the dynamism of human creativity. It is a 'freezing of life', he argues, that 'might be described in Hegelian terms, as an absolute predominance of "tranquil side-by-sideness" in space over 'restless becoming in the progression of time' (1994: thesis 170). Other situationists had also long been concerned with urbanism's role in congealing space according to hierarchical social interests. Writing on behalf of the situationist 'Bureau of Unitary Urbanism' in Brussels in 1961, Attila Kotányi and Raoul Vaneigem (1981: 67) refer to urban planning as a 'geology of lies'. For them urbanism is a form of spectacular ideology that represents the materialization of particular capitalist interests to the exclusion of others, and that has even reshaped basic spatial laws so as to hide those partial interests behind claims to speak for the whole. Urbanism works as a kind of propaganda, insisting on notions of public participation and popular consent and smoothing the wheels of social subjection through its promotion of a supposedly conflict-free image of order and rational organisation. Vaneigem (1996: 120) developed a vigorous attack on urbanism elsewhere in the situationist journal, describing contemporary cities as 'fossilized', and asserting that there is 'an incredible leadenness [*lourdeur*] in everything related to urbanism'. This weighty terrain represents a channelling of desire and a tangible form of alienation in itself. Its ultimate dream, according to Vaneigem, is that of a world in which human beings have become assimilated to concrete.

For the situationists, then, the language of freedom and choice presented by the spectacle cloaks the capitalist production of a space of *un*freedom and constraint. In many of their writings, urban planning is presented as the enemy of the city as a realm of possibility and emancipation. In such critiques, the situationists frequently concentrate their barbed written assaults on the large-scale restructuring of city centres such as Paris during the 1950s and 1960s, and especially the rapid construction of new towns and suburban housing projects during the same period such as those in France at Sarcelles and Mourenx – towns they denounce as 'laboratories of this stifling society' (SI, 1989a [1964]: 161). To this end they often hold up the promises of capitalist urbanism – with their claims about freedom, liberation and happiness – as measures against which to address the shortcomings of urban life and its environments as actually produced. They suggest that the billboards promising happiness through the purchase of the latest commodity and housing unit, or media announcements of a new so-called 'leisure society' and 'liberation from work' transforming life in the city, should be taken at their word – and their objects of promotion judged accordingly. The language of spectacular urbanism should be confronted and turned around, and desires that are currently buried in inert forms of urbanism should be uncovered and rejuvenated, through the contestation of what exists and attempts to imagine and produce alternatives.

Moving situations

The second quotation from Debord at the start of this chapter refers to the SI's aim to contest dominant urban conditions and to develop contrasting visions of the city. It comes from a key early text that was presented as a discussion document at the foundation of the group in July 1957. Debord here underlines his commitment to revolutionary action within culture, arguing that its aim must be not to translate or explain life but to enlarge it. A starting point lies in working with everyday approaches and conditions that exist, while recognizing that they must be criticized and superseded in any genuinely critical endeavour. In a section entitled 'Towards a Situationist International', he outlines provisional means by which the situationists can pursue his interventionist strategy and calls for the invention of new types of games as a way of expanding life and enhancing freedom. 'The situationist game is distinguished from the classic conception of games by its radical negation of the element of competition and of separation from everyday life,' he states. 'On the other hand, it is not distinct from a moral choice, since it implies taking a stand in favour of what will bring about the future reign of freedom and play' (Debord, 1985: 617; tr. modified from 1981b: 24).

Games had featured strongly in attempts by forerunners of the SI, the Letterist International (LI),[1] to intervene in urban space and disrupt the 'frozen' terrain of contemporary urbanism. Since the early 1950s this small group of activists and anti-artists, who counted Debord among their number, had explored the freedoms of cities and especially Paris where they were based. Much of their activity went under the sign of 'psychogeography', which was concerned with the interactions between urban environments and behaviour. They reported on urban ambiences, atmospheres, and the emotional contours of cities as discovered on critical drifts or *dérives*. At the same time they contested constraints and challenged conventional understandings of cities. The letterists opposed the limitations set around play in cities and, through practice of the *dérive*, they not only studied urban spaces, including the potential for play within them, but also participated in a game within those spaces based on a more liberated way of life that they sought to extend through experiment and constructive practices. In this sense the drift was not simply a technique of observation but a passional journey in itself that worked against the petrifaction the LI associated with contemporary urbanism. It sought instead to chart other paths while attending to how cities could be changed and alternative spaces constructed. Indeed, the term *dériver* itself suggests an active sense of untying binds and undermining fixity. The term carries watery connotations of drifting that are prominent in common understandings of the practice, with its roots coming from *rivus* or stream. At the same time it is inflected by the English word to drive, with its sense of intention. It further contains within itself, as the letterists themselves noted, the sense of '[to] undo what is riveted [*rivé*]' (LI, 1996 [1956]: 60). Behind it is the dream of undoing the 'geology of lies' of urban planning and of people exploring the freedoms of urban life and creativity themselves.

Debord refers to such psychogeographical practices in his text of 1957, placing them at the heart of the new situationist agenda. Debord (1985: 617; tr. 1981b: 24) views the *dérive* as an approximate experimentation towards a new mode of behaviour and suggests that such a 'will to playful creation must be extended to all known forms of human relationships'. But the central idea that brings these practices together politically is 'the construction of situations'. The phrase again came out of the LI's activities. Announcing their intention 'to establish an exciting way of life', members of that group once stated: 'We experiment with behaviour, decoration, architecture, urbanism and communication meant to create attractive *situations*' (Bernstein et al., 1996 [1954]: 50, emphasis in the original). As part of their critique of everyday life, the letterists and later the situationists used the idea to suggest that people's lives could be understood as a succession of situations. To intervene politically it was necessary to challenge the everydayness and routinisation of situations, with their accompanying sense of limited possibility, and to construct situations actively and consciously.

The situationists differentiated their approach from academic theorizing by emphasizing the sense of revolutionary intention and activism that lay behind it. They were concerned not simply with studying the situations within which human subjects find themselves, but with consciously making new situations and transforming the conditions of everyday life. Debord describes the spatio-temporal practice of constructing situations as 'the concrete construction of momentary ambiences of life and their transformation into a superior passional quality'. This entails 'a methodical intervention based on the complex factors of two components in perpetual interaction: the material environment of life and the comportments which it gives rise to and which radically transform it' (Debord, 1985: 616; tr. 1981b: 22). The construction of situations for the SI is direct, intense, emotionally charged, and participatory. Its interventionist and lived nature means that it is by definition opposed to the principles of the spectacle, beginning on its 'ruins'. It reaches towards what Debord (1997: 90) refers to elsewhere as 'immediate participation in a passionate abundance of life, through the variation of fleeting moments resolutely arranged'.

In contrast to the frozen qualities of contemporary urbanism, the situationists understood the construction of situations to be a continuous process characterised by the provisional and fleeting. They scorned notions of permanence and eternity as goals of aesthetic production. Distancing themselves from 'aesthetic modes that strive to fix an emotion', they sought to push 'ever further the game of creating new, emotionally moving situations' (Debord, 1985: 618). The situationist approach was experimental and aimed to open onto the unknown and yet to be discovered. They recognised that appropriate techniques had still to be invented. Initial plans would necessarily prove inadequate, and working hypotheses would have to be continually re-examined. But they nevertheless believed that material conditions supported the salience of their approach. Debord insisted:

We have to multiply poetic subjects and objects – which are now unfortunately so rare that the slightest ones take on an exaggerated emotional importance – and we have to organize games for these poetic subjects to play with these poetic objects. This is our entire program, which is essentially transitory. Our situations will be ephemeral, without a future. Passageways. (1985: 618; tr. 1981b: 25)

A similar striving to shatter the constraints of social life and subvert current conditions runs through situationist writings. Echoing a metaphor that has been threading through this chapter, Vaneigem later stated that the group was fundamentally concerned only with 'the moments when life breaks through the glaciation of survival' (1981 [1963]: 122). And addressing questions about the city more directly, he argued with Kotányi that the process of countering current urban conditioning is already the construction of situations, something they defined as 'the liberation of the inexhaustible energies trapped in a petrified daily life' (Kotányi and Vaneigem, 1981 [1961]: 67). But how could such liberation change urban space fundamentally? How could it go beyond small-scale, momentary interventions to produce a new social space? How might an 'organized collective work' to revolutionize everyday life, of the kind demanded by Debord, enable the construction of a different urbanism for a different life? These questions were of key importance in the SI's moves towards what it called 'unitary urbanism', to which I now turn.

Unsettling urbanism

Unitary urbanism became a central concern in the SI's early years as part of its attempts to transform social space and urbanism. It provided a framework within which to develop ideas about the construction of situations and the extension of their influence in space and time. The term was first used towards the end of 1956 in meetings that led to the foundation of the group, although it had effectively emerged as a concept through discussions three years before. Debord (1985: 616; tr. 1981: 22) gave the idea an experimental but essential position in the SI's agenda in 1957 when he suggested that unitary urbanism could be defined 'first of all by the use of the ensemble of arts and technics as means contributing to an integral composition of a milieu'. In keeping with discussions with colleagues at the time, he recognized the important role of architecture in this construction. But he eschewed formalist emphasis on architectural or urbanist innovation and instead focused attention on the 'architectural complex' as the basic unit of construction through which diverse means and events – among them sound, art, cinema, poetry – could be brought together and used to condition ambiences and produce effects at the level of the situation. He argued that architecture 'must advance by taking emotionally moving situations, rather than emotionally moving forms, as the material it works with' (1985: 617; 1981b: 23).

Unitary urbanism was therefore not a doctrine, the preserve of a specialist discipline awaiting technical implementation. Instead, it was a critique of urbanism. Debord and the situationists insisted that it came out of current struggle and experimentation, already having an active presence through its research and development. At the same time, it had a visionary element.

Unitary urbanism took its cues from the possibilities in cities and everyday life that were as yet unrealized, with the SI (1996: 112) wanting it to function as 'a hypothesis for using the means available to humanity today to freely construct its life, beginning with the urban environment'. As a vision of urban space, it opposed the stasis and techniques of separation associated with dominant forms of urbanism, and it attacked the roles of commercial interests and the forces of capital investment as well as mainstream planning practice in defining spaces in the city. While urban planning at the time was strongly influenced by modernist doctrine on the need to divide functions strictly into separate zones, including those of work and residence, Debord asserted in contrast that it was important to 'at least envisage a third sphere: that of life itself (the sphere of freedom, of leisure – the truth of life)'. He insisted that unitary urbanism 'acknowledges no boundaries' and that 'it aims to form a unitary urban milieu in which separations such as work/leisure or public/ private will finally be dissolved'. The goal of revolutionary urbanists was to break topological chains and, in so doing, to pave 'the way with their experiments for a human journey through authentic life' (1981c: 57, 58).

Unitary urbanism therefore reached beyond particular issues and problems such as those concerning housing or aesthetics, to look towards a general construction of social space that transcended segmentations and that was dynamic in all respects. It was an attempt to envisage a space constructed according to the interests and desires of its inhabitants, no longer alienated from their surroundings and their true selves, no longer isolated from one another, freely creating their environments in tune with their fluctuating activities as an ever-evolving work or *oeuvre*. In opposition to the spectacle, it was meant to be collective and participatory. And attempting to go beyond questions of utility and the world of labour, it advocated realizing the potential for a playful environment that would be directly lived by its creators. As the situationists put it, 'unitary urbanism envisages the urban environment as the terrain of a game in which one participates'; it is 'the foundation for a civilisation of leisure and of play' (SI, 1989b: 144).

Difficulties occupied much of this discussion of unitary urbanism, however. Many of these are common to emancipatory visions of urbanism more generally. Among them are: how can such urban spaces be imagined and created in ways that do not themselves reproduce the restrictions of those they seek to surpass? How can they avoid settling into forms that deny other freedoms and aspirations, given the ways in which they embody particular interests and desires? How can they resist becoming compensatory distractions or even authoritarian forms that constrain and control? Utopian thought about cities is replete with examples of visions of freedom that become prison-like in their construction as they seek to fix history and freeze geography in an ideal form. The classic blueprint model of utopian urbanism, which is typically based on an ordered spatial form, aims to provide the means for overcoming urban ills and establishing an ordered, harmonious society. But within such perspectives uncertainty, chance, and even change itself become construed as threatening to the aim of realizing the blueprint. What does not 'fit' with such formal plans becomes a 'problem', the waste productions of spatial ordering and purification. A further tension appears in that the imaginings of alternative futures have to take place under the restricted conditions of the present that such imaginings would seek to abolish. Yet the desires and needs that would ultimately shape the new city cannot be known in advance of the radical change and emancipation that would allow them to emerge.

The situationists, however, avoided blueprint models and ordered spatial forms in their accounts of unitary urbanism. Their emphasis was on fluidity and flexibility. They argued that unitary urbanism opposed the temporal fixation of cities, favouring perpetual transformation and movement. They also believed that such dynamism would extend to the everyday lives of inhabitants by opposing the fixation of people at certain places in the city. Construction would be a continual process with some elements being new, others being appropriated, diverted, and reworked from existing cities by means of what the SI called *détournement* (SI, 1989b [1959]). The group even wrote of cities being constructed for perpetual *dérives*. Far from seeking to establish a settled ideal, then, such aims and tactics suggest something more akin to a 'vagabond' urbanism that is transgressive, oppositional, and nomadic.[2] The restlessness of the situationist approach may be seen as a deliberate attempt to unsettle, to disrupt the passifying conditioning of contemporary spectacular urbanism as well as traditional compensatory fantasies, which utopian thought often associated with good form. But the approach also connects to their central claim: that unitary urbanism would always be in the process of becoming, the collective creation of its inhabitants through the struggle for change. It could not be prescribed in advance. It would be produced through that 'liberation of inexhaustible energies' in daily life and the transformation of both space and society. It might be noted that this approach does not fit easily with the desire in some recent writings on the situationists to emphasise their architectural interests, especially where that involves attempting to extract an 'architectural theory' and 'design theory' from their revolutionary programme (for example Sadler, 1998).

Tensions within the concept of unitary urbanism nonetheless remained, as are vividly apparent in the most fully realized attempt to envisage aspects of unitary urbanism from within the situationist movement, one produced by the Dutch artist Constant. Working initially as a member of the SI before resigning in 1960 and continuing in the subsequent decade outside the group, he embarked on a remarkable series of models, writings, art works, lectures, and the like through which he outlined a new urban culture and space entitled 'New Babylon'. Underpinning his works is a dream of the liberation of humanity and the flowering of the potential for play and the free creation of space, enabled by the automation of productive labour and the collective socialisation of land. He attempted to encourage a different way of thinking about cities and urban life based on this interest in liberation and play. His urban spaces encapsulate the concern with openness, flexibility, fluidity, and provocation discussed above.

Writing with Dutch colleagues in the SI, he suggested that the great strength of the contemporary artist lay in the 'acceptance of the transitory' in contrast to those individual arts that are 'tied to an idealist conception, to a seeking after the eternal'. They stated:

> By renouncing fixed form, we arrive at all forms, which we invent and afterwards reject ... It is uninterrupted invention that interests us: invention as a way of life ... Only urbanism will be able to become that unitary art which responds to the exigencies of dynamic creativity, the creativity of life. Unitary urbanism will be the ever variable, ever alive, ever actual, ever creative activity of the man [and woman] of tomorrow. (Alberts et al., 1996 [1959]: 90)

Constant's projects do not shy away from the challenge of addressing what this 'ever variable' and 'ever creative' activity might involve. The shifting perspectives and disruptive energy of his models and art works, with their emphasis on movement and change, are not only critiques of contemporary urbanism. They are also attempts to imagine alternatives while recognizing the difficulties involved. With their tensions and at times disturbing edge they are testament to Constant's attempts to negotiate the paradoxes involved in unitary urbanism, in recognition that utopian schemes all too often become repressive and constraining, and that ultimately the builders of this city space will not be planners or visionaries but – as he often emphasized – the New Babylonians themselves (for further discussion, see Pinder, 2001, 2005).

Conclusions

In this chapter I have traced a line through certain situationist attitudes towards the city and emancipatory spaces, one that has centred on the group's ambivalent relationship with cities especially in terms of the dialectics of fixity/fluidity, freezing/flowing, stasis/life that run through their approach. It should be clear that the situationists' attacks on the congealed qualities of the urban, and their emphasis on flexibility and fluidity, were not simply matters of formal architecture or design. They offered a *critique* of geography that understood space as socially produced, as they sought a lived space that would be dynamic and mobile in that it would involve the 'ever variable', 'ever alive' and 'ever creative' activity of socio-spatial change. It would be a space inherently at odds with current urbanism's 'geology of lies', being produced through challenging alienated urban existence and consciously constructing situations and spaces in ways that would allow the expression and realization of the desires of urban dwellers.

It should be noted that attitudes towards unitary urbanism varied greatly within the SI and that they also changed markedly over the years of the group's existence. Between Debord's report at the foundation of the SI in 1957 and the publication of his *The Society of the Spectacle* ten years later, the term 'unitary urbanism' was dropped from the group's vocabulary. The emphasis moved to political critique of existing conditions. Urban interventions focused on reappropriating spaces and making guerrilla raids to subvert conditions and contest society as a whole. The tactics remained mobile, operating in the space of the other through means of *détournement*. Debord (1994: theses 206, 208) presented the latter practice as a device for restoring all the 'subversive qualities to past critical judgements that have congealed into respectable truths', and as 'the fluid language of anti-ideology'. However, there were relatively few situationist attempts to envisage the production of alternative urban spaces along the lines of the earlier experiments.

Indeed many situationist texts from the early 1960s convey the impression that developing urban visions within the present, prior to the revolution of everyday life that the group aimed to encourage, would necessarily fall into the trap of urbanistic ideologies. Histories of the group typically make much of this shift, representing it as part of a general rejection of earlier 'artistic' and urban-centred positions in favour of a more explicitly political-theoretical stance. Yet it is important to consider how questions of urban space remained crucial within its political programme through the 1960s. Debord tellingly refers to the process of revolution in *The Society of the Spectacle* as a '*critique of human geography*' (emphasis in the original) through which people make their own spaces and histories. The lines following this phrase further take up themes that have run through the discussion of unitary urbanism and the construction of situations in this chapter. 'By virtue of the resulting mobile space of play, and by virtue of freely chosen variations in the rules of the game,' Debord wrote, 'the independence of places will be rediscovered without any new exclusive tie to the soil, and thus too the authentic *journey* will be restored to us, along with authentic life understood as a journey containing its whole meaning within itself' (1994: thesis 178, emphasis in the original). His words would be given particular force by the revolutionary events the next year. Less than six months after his book was published, the 'explosion' in Paris in May 1968 saw urban spaces appropriated and transformed through many of the tactics Debord had envisaged with the concept of unitary urbanism.

In an era that badly needs to challenge the complacencies of current urban thinking, it is worth returning to debates about unitary urbanism not only for what they reveal about the situationist programme itself, but also as part of a reappraisal of critical approaches to cities (Baeten, 2002; Harvey, 2000). Avant-garde positions like the unitary urbanism explored by the situationists may seem out of time in many respects, but the questions they raise about representation, imagination, participation, and pleasure are critical in expanding our senses of urban politics and its possibilities. Their experiments with movement and the construction of space-times connect with demands for 'rights to the city' and with interest in cities as sites of politics in motion and for 'testing new ground' (Amin and Thrift, 2002; Lefebvre, 1996). Situationist dreams of urban emancipation may not have been realized but elements can still play a role in fueling subversive energies in the future. As new challenges to unequal power relations emerge in cities around the world, and as people reappropriate and reconstruct urban spaces with the aims of social transformation, Debord's argument about cities taking possession of the freedoms that have hitherto been denied them may yet find its time.

Notes

1 I am here adopting the Anglicisation of the original French name *Internationale Lettriste* that was favoured by the group itself, as Andrew Hussey (2001: 377) points out. The LI was formed in 1952 as a radical splinter from the Letterist Movement, and it was one of the groups involved in the meetings in Italy that led to the foundation of the SI in July 1957. The other main group was the International Movement for an Imaginist Bauhaus.
2 I take the term from Anthony Vidler's (1992: 207–14) discussion of a 'vagabond architecture', which centres on the work of John Hejduk but includes references back to the SI.

References

Alberts, A., Armando, Constant and Oudejans, H. (1996) 'First proclamation of the Dutch section of the SI', originally published 1959, trans. P. Hammond, in L. Andreotti and X. Costa (eds), *Theory of the Dérive and Other Situationist Writings on the City*. Barcelona: Museu d'Art Contemporani de Barcelona and Actar. pp. 89–90.

Amin, A. and Thrift, N. (2002) *Cities: Reimagining the Urban*. Cambridge: Polity.

Baeten, G. (ed.) (2002) 'The spaces of utopia and dystopia', Special issue, *Geografiska Annaler*, 84B: 3–241.

Bernstein, M., Dahou, M., Vera and Wolman, G. (1996) 'The general line', originally published 1954, trans. G. Denís, in L. Andreotti and X. Costa (eds), *Theory of the Dérive and Other Situationist Writings on the City*. Barcelona: Museu d'Art Contemporani de Barcelona and Actar. p. 50.

Debord, G. (1967) *La société du spectacle*. Paris: Buchet-Chastel.

Debord, G. (1981a) 'Theory of the dérive', originally published 1956, in K. Knabb (ed.), *Situationist International Anthology*. Berkeley, CA: Bureau of Public Secrets. pp. 50–4.

Debord, G. (1981b) 'Report on the construction of situations and on the International Situationist Tendency's conditions of organization and action' (excerpts), in K. Knabb (ed.), *Situationist International Anthology*. Berkeley, CA: Bureau of Public Secrets. pp. 17–25.

Debord, G. (1981c) 'Situationist theses on traffic', originally published 1959, in K. Knabb (ed.), *Situationist International Anthology*. Berkeley, CA: Bureau of Public Secrets. pp. 56–8.

Debord, G. (1985) 'Rapport sur la construction des situations et sur les conditions de l'organisation et de l'action de la tendance situationniste internationale', originally written 1957, reprinted in G. Berreby (ed.), *Documents rélatifs à la fondation de l'Internationale situationniste*. Paris: Editions Allia. pp. 607–19.

Debord, G. (1992) 'On the passage of a few persons through a rather brief period of time', originally published 1978, trans. K. Knabb, in G. Debord, *Society of the Spectacle and Other Films*. London: Rebel Press.

Debord, G. (1994) *The Society of the Spectacle*, trans. D. Nicholson-Smith. New York: Zone Books.

Debord, G. (1997) 'Theses on the cultural revolution', originally published 1958, trans. J. Shepley, in T. McDonough (ed.), Guy Debord and the *Internationale situationniste*, *October*, 79: 90–2.

Harvey, D. (2000) *Spaces of Hope*. Edinburgh: Edinburgh University Press.

Hussey, A. (2001) *The Game of War: The Life and Death of Guy Debord*. London: Jonathan Cape.

Kotányi, A. and Vaneigem, R. (1981) 'Elementary program of the Bureau of Unitary Urbanism', originally published 1961, in K. Knabb (ed.), *Situationist International Anthology*. Berkeley, CA: Bureau of Public Secrets. pp. 65–7.

Lefebvre, H. (1996) 'The right to the city', originally published in 1968, in E. Kofman and E. Lebas (trans. and eds), *Writings on Cities*. Oxford: Blackwell. pp. 61–181.

LI (1996) 'Towards a lettrist lexicon', originally published 1956, trans. G. Denís, in L. Andreotti and X. Costa (eds), *Theory of the Dérive and Other Situationist Writings on the City*. Barcelona: Museu d'Art Contemporani de Barcelona and Actar. p. 60.

Pinder, D. (2000) '"Old Paris is no more": geographies of spectacle and anti-spectacle', *Antipode*, 32: 357–86.

Pinder, D. (2001) 'Utopian transfiguration: the other spaces of New Babylon', in I. Borden and S. McCreedy (eds), *New Babylonians*, theme issue of *Architectural Design* 71: 15–19.

Pinder, D. (2002) 'In defence of utopian urbanism: imagining cities after the "end of utopia"', *Geografiska Annaler* 84B: 229–41.

Pinder, D. (2005) *Visions of the City: Utopianism, Power and Politics in Twentieth-Century Urbanism*. Edinburgh: Edinburgh University Press.

Sadler, S. (1998) *The Situationist City*. Cambridge, MA: MIT Press.

SI (1981) 'Questionnaire', originally published 1964, in K. Knabb (ed.), *Situationist International Anthology*. Berkeley, CA: Bureau of Public Secrets. pp. 138–42.

SI (1989a) 'The world of which we speak', originally published 1964, trans. T. Levin, in E. Sussman (ed.), *On the Passage of a Few People Through a Rather Brief Moment in Time: The Situationist International, 1957–1972*. Cambridge, MA: MIT Press. pp. 154–73.

SI (1989b) 'Unitary urbanism at the end of the 1950s', originally published 1959, trans. T. Levin, in E. Sussman (ed.), *On the Passage of a Few People Through a*

Rather Brief Moment in Time: The Situationist International, 1957–1972. Cambridge, MA: MIT Press. pp. 143–7.

SI (1996) 'Critique of urbanism', originally published 1961, trans. P. Hammond, in L. Andreotti and X. Costa (eds), *Theory of the Dérive and Other Situationist Writings on the City.* Barcelona: Museu d'Art Contemporani de Barcelona and Actar. pp. 109–15.

Vaneigem, R. (1981) 'Basic banalities (II)', originally published 1963, in K. Knabb (ed.), *Situationist International Anthology.* Berkeley, CA: Bureau of Public Secrets. pp.118–33.

Vaneigem, R. (1996) 'Commentaries against urbanism', originally published 1961, trans. P. Hammond, in L. Andreotti and X. Costa (eds), *Theory of the Dérive and Other Situationist Writings on the City.* Barcelona: Museu d'Art Contemporani de Barcelona and Actar. pp. 119–24.

Vidler, A. (1992) *The Architectural Uncanny.* Cambridge, MA: MIT Press.

8 Everyday Rationality and the Emancipatory City

Gary Bridge

The act of walking is to the urban system what the speech act is to language or the statements uttered. (de Certeau, 1984: 97)

In the above quotation Michel de Certeau (1984) alludes to the creative and innovative possibilities of walking the city. Speech acts often play on the proper use of language to make new meanings, and in a similar way walking can play on the proper meaning of the city. A walk can thread together diverse locations and situations and otherwise disrupt the proper geography of the city dictated by the rational plan. De Certeau sees in the act of walking the possibility to experience the city differently: in the same way that 'speech acts' can change inherited meanings in language. Pedestrian wanderings in the city can help give a feeling of personal freedom that can be experienced in secret from the gaze of urban authorities and the urban order: 'The user of the city picks out certain fragments of the statement in order to actualize them in secret' (1984: 98). The effect here is presumably a sense of personal liberation, of being the author of one's own urban story.

The figure that most emphatically combines stylistic innovation with a secretive existence in the city is the flâneur. This male, moneyed, non-working, fashion-conscious urbanite moved freely through the streets of nineteenth century Paris (and other European capitals) observing the lives of others in a passive but acquisitive way: adding the myriad of urban stories to his own sense of growing cosmopolitan experience. De Certeau acknowledges that not all urban dwellers can move as freely as the male and moneyed flâneur but suggests, tantalizingly, that all urban dwellers have the potential to impress their own pedestrian story on certain parts of the city. Urban practices such as walking can be a form of individual window-shopping on the urban experience.

The flâneur's freedom to roam the city and garner the rich experiences of urban life is a form of emancipation. The city, through the anonymity it affords, gives a space for personal freedom and the development of a cosmopolitan personality. But this is a bourgeois, private, individualized and selfish sense of emancipation. The flâneur is a voyeur, watching but not speaking with others.

The example of the flâneur highlights the point of de Certeau's speech act metaphor. It shows that style alone is not enough. In order to succeed innovative speech acts must be received and acknowledged by an audience. To do that they must conform to a norm, lest the unusual speech act be interpreted as an act of strangeness while still being innovative enough to alter the proper meaning and enable a different intent to be understood.

The ways that new speech acts might succeed in being acknowledged by an audience, in 'coming off', as John Austin (1962) puts it, is a matter of some debate. For Jürgen Habermas, as we shall see, it is the exchange of meaning guided by rational procedures of debate that helps secure a new consensus. For Richard Sennett it is stylized, dramaturgical elements of communication cutting across social difference that give speech acts credibility with an audience and help new statements to be made. The problem is that Habermas's communicative rationality initially requires exclusive communities of communication to get the debate going. Similarly, Sennett's dramaturgical solution depends on certain exclusive spaces of communication in the city, such as the coffee house in the eighteenth century city. In this chapter I argue that extending communication beyond small communities and exclusive spaces relies on a form of strategic rationality – a set of rational expectations between the speaker and the audience – to enable the speech act to come off. The paradox is that such strategic rationality is traditionally seen as an oppressive force in urban social relations, encouraging selfish, non-trusting behaviour. Despite this, I shall argue that strategic rationality is at its height in the public spaces of the city and essential to any wider emancipatory aims.

The perils of rationality

Critical theorists, even those, such as Michel Foucault (1986), who point to the historical-geographical particularities of rationality, do not deny its pervasiveness in everyday life. Indeed for Max Horkheimer and Theodor Adorno (1986) and other members of the Frankfurt School instrumental rationality has surpassed exploitation of labour as the most pervasive source of domination in modern societies. That claim was anticipated to some extent by Georg Simmel (1995) and his recognition that the new types of relationships he saw unfolding in the streets of turn-of-the-century Berlin were predicated on people's intellectual reactions to each other. In situations where passers-by are unknown to each other and do not share a common history, cognitive functions take over from custom or tradition in guiding the anticipations people have of each other. As a result individuals attempt to limit the amount of information they give off about themselves in public, hiding behind the mask of rationality and displaying an indifferent attitude to each other and to the overstimulation of the city. Simmel thought urban social relationships were like monetary exchanges in their degree of disinterestedness. The deadening and alienating conditions of interaction leave urban dwellers prone to the wider rationalities of the market and bureaucracy.

For de Certeau (1984) the rationalities of the city work according to a 'scopic regime', much like the way that Foucault (1977) described panopticon. De Certeau argues that this scopic regime is most clearly manifest in the very concept of the city as a totality. The 'concept city' is characterized by the production of its own space (repressing other spatial practices), a synchronic system of power (a 'nowhen') that overrides local traditions, and the establishment of the city itself as a universal and anonymous political entity. Instrumental rationality is hegemonic in the concept city. The city as totality is reflected in the rational plan, a view from above the city that seeks control by treating it as a machine or a human body. The heteronymous practices of city dwellers are ignored in this scheme. City dwellers themselves are treated as objects in the means-ends rationality of the master plan and utilitarian bureaucracy, as numbers in the professional discourses of social administration. The city itself is objectified as a series of rational, functional spaces that enable the economic system as a whole to function smoothly.

Yet for de Certeau there are certain urban practices that can act as forms of resistance against this dominating logic. The most prominent is the act of walking the city. 'The walking of passers-by offers a series of turns (tours) and detours that can be compared to "turns of phrase" or "stylistic figures". There is a rhetoric of walking. The art of "turning" phrases finds an equivalent in an art of composing a path (*tourner un parcours*)' (1984: 100). This pedestrian rhetoric works on and against 'the geometrical space of urbanists and architects' which 'seems to have the status of "proper meaning"' (p. 100). So just as new turns of phrase can add novelty to language, new paths and styles of walking the city can work against the rationalist logic of the planners. They can create a labyrinth of connotations and associations that are the creative outcomes of urban practices, rather than the sleepwalking induced by the modernist master plan. Certain pedestrian experiences can come to represent larger stories in the city more widely, just as in language synecdoche is the naming of the part of an object that comes to represent the whole object (sails for ships, for example). The urban pedestrian can also take shortcuts, or splice together routes and experiences that the master logic of the concept city normally keeps apart, just as in language asyndeton suppresses conjunctions and adverbs. This can result in striking statements and juxtapositions. A slight change, such as the omission of a conjunction, can radically alter the sense of what is said (compare stop start with stop and start, for example).

Emancipation and enunciation

De Certeau realizes the essentially personalized and cognitive emancipations that may result from walking the city: what he calls a 'discreteness' of the experience of cities. The city walker:

> creates a discreteness, whether by making choices among the signifiers of the spatial 'language' or by displacing them through the use he makes of them. He condemns certain places to inertia or disappearance and composes with others spatial 'turns of phrase' that are 'rare', 'accidental' or 'illegitimate'. (1984: 98–9)

These individualized and stylistic renderings of the city are reminiscent of flânerie. If we follow the metaphor of the speech act then stylized gestures will not 'come off' unless they are received and acknowledged by an audience. Intriguingly, de Certeau expresses the linguistic relationship between speaker and audience as the equivalent of a form of bodily location in the city:

> In the framework of enunciation, the walker constitutes, in relation to his position, both a near and a far, a here and a there. To the fact that the adverbs here and there are indicators of the locutionary seat in verbal communication – a coincidence that reinforces the parallelism between linguistic and pedestrian enunciation – we must add that this location (here-there) (necessarily implied in walking and indicative of a present appropriation of space by an 'I') also has the function of introducing an other in relation to this 'I' and of thus establishing a conjunctive and disjunctive articulation of places. I would stress particularly the 'phatic' aspect, by which I mean the function, isolated by Malinowski and Jackobson, of terms that initiate, maintain, or interrupt contact, such as 'hello', 'well, well' etc. Walking, which alternately follows a path and has followers, creates a mobile organicity in the environment, a series of phatic topoi. And if it is true that the phatic function, which is an effort to ensure communication, is already characteristic of the language of talking birds, just as constitutes the 'first verbal function acquired by children', it is not surprising that it also gambols, goes on all fours, dances, and walks about, with a light or heavy step, like a series of 'hellos' in an echoing labyrinth, anterior or parallel to informative speech. (1984: 99)

I think that there is much here that speaks to the idea of the emancipatory city. The sense of location in the city is a function of the range of enunciations that can be made, moved, and put into contact with other enunciations. The ability or inability to 'establish a conjunctive and disjunctive articulation of places' can relate not just to location but also to the locatedness of speakers. For me the idea of locatedness is the ease of being in the city and the ability to initiate and splice together different enunciations. In this sense 'locatedness' offers a more promising and emancipatory refuge from the distanciation and disorientation of the modern city than the notion of 'belonging', with all its dangers of a reactionary politics of essentialism, blood, and soil. Alternatively Frederic Jameson (1984) uses Kevin Lynch (1960) and his idea of cognitive legibility to try to establish orientation in the city. The benchmark against which Lynch measures how people mentally map the city is the ability to visualize the whole city in the way a planner would. This ties it to the overall logic of the concept city and its oppressive imagination of the visual.

These ideas capture a wider tension between spatial order and social process that David Harvey (2000: 133–96) identifies in his discussion of the spaces of utopia. Harvey at one point criticizes Foucault's notion of heterotopia because the sorts of spaces that may allow 'otherness' to be established (such as colonies, prisons and brothels) cannot confine social processes to acts of resistance or alterity. These same social processes might result in banal or exclusionary practices that domesticate the emancipatory potential of the oppositional space. Harvey turns to the idea of socio-temporal utopianism in response to the mismatch of previous spatial orders and social processes. I agree with this but would argue that a sense of location (and its emancipatory possibilities) comes out of mobility, in space and language, that avoids the more onerous communication found in a settled place or community.

I would adapt de Certeau to form a notion of enunciatory locatedness or what we might call a 'speech place'. A successful speech act uses mobility in language to bring together different utterances in novel ways and out of that to establish the presence and effect (or subjectivity) of the speaker. This happens through the ability of neologisms and novel thoughts to make speech 'act' on the hearer. In a similar way movement through the city and engagement with different social situations gives pedestrians a spatial lexicon that enables them to draw the city together in different ways. This feeling of location or 'speech place' comes from the ability to move easily in and out of different social communications and contexts and connect with different audiences. A speech place is established by phatic utterances: the use of speech to establish, maintain, interrupt, and break off communication – the 'hellos' and 'well wells', as de Certeau describes it.

De Certeau's metaphor of speech acts and pedestrian rhetoric thus takes us beyond the private city of the flâneur. He distinguishes between 'style', which registers symbolically in language an individual's way of being in the world (like the style of the flâneur), and 'use', which refers to the social norms that support the intelligibility of communication. 'Style' and 'use' both have to do with a way of operating (of speaking, walking, etc.), but style involves a peculiar processing of the symbolic, while use refers to elements of a code. They intersect to form 'a style of use, a way of being and a way of operating' (1984: 100). The other element of speech acts is that to have illocutory force they must resonate with an audience, and to do *that* they must operate within existing norms and conventions. Otherwise the speech act will 'misfire' (Austin, 1962). Speech acts only come off within conventions of acceptance and interpretation.

So far the elements of pedestrian rhetoric have been used to suggest improvisations that work around the prevailing rationalist logic or 'proper meaning' of the city. These improvisations occur most readily at the edges of communication, in what de Certeau (1984: 99) calls the phatic function of language ' establishing, maintaining, and terminating dialogue. As Erving Goffman (1959) and others working in the social/symbolic interactionist tradition of sociology have argued, the fringes of interaction, especially with strangers, are governed as much by rational strategy and by participants' pre-suppositions about the presuppositions of others as they are by dramaturgical expression involving conversational routines, forms of body presentation, and interaction rituals. Goffman acknowledges the contextual (or indexical) importance for language but adds:

> Whatever else, our activity must be addressed to the other's mind, that is, to the other's capacity to read our words and actions for evidence of our feelings, thoughts and intent. This confines what we say and what we do, but it also allows us to bring to bear all the world to which the other can catch allusions. (1983: 50–1)

This is a shared imaginary realm but one that is crucial in establishing the reality of everyday life. In his recognition of the strategic aspects of social interaction, Goffman was drawing on the work of the game theorist Thomas Schelling. Schelling (1960) argued that communicative interaction was guided by the rational expectations of players. At the margins of communication involving phatic utterances and body language, strategic rationality is doubly implicated: in both governing the interaction routines and the first-impression assessments of strangers entering the interaction. Yet it is also at these margins where innovation in speech (through novel speech acts) is most likely. To succeed, any innovatory speech acts will have to conform or at least resemble some norm or rational expectations (Bridge, 1997; 2000). There is good reason to believe that the norms that emerge in new situations of inter-action are in fact coordinations of rational expectations. Robert Sugden (1989) gives the famous example of a set of social norms arising over the right to collect driftwood from a beach. New situations of interaction need rational coordination in the absence of pre-existing traditions or customs.

What all this suggests for those speech acts is that to succeed they must conform to a 'use', or norm, based on mutually recognized (however tacitly) rational expectations between speaker and audience. 'Style' must indeed combine with 'use' to make the innovatory 'act' be appreciated, just as intention must conform to convention to make the speech act come off. In the same way the walker in the city has the potential to innovate but must do so in a way that meets the rational expectations of each urban situation she encounters.

At this point in the argument we are caught by what might be called Simmel's revenge. Simmel's argument is that in the modern metropolis the prevailing convention of interaction is one of indifference, involving minimized and neutral exchanges of information among strangers. It is unlikely that innovative signals in communication could ever come off in this prevailing atmosphere. In New York, London, and other cities passengers crowd into commuter trains, pressing up against each other in silence with their faces buried in newspapers averting each others' gazes. Mutual suspicion and self-repression behind the mask of rationality make the modern city very unpromising ground either for giving off or receiving innovative signals.

What I am suggesting in this chapter is that even the most pared down, tacit forms of communication work to the same structure of rational expectation and interpretation underpinning much richer forms of communication. Fleeting moments of tacit communication can play off ratified forms of communication (the proper language for speech acts, the proper meaning of the concept city) because they are both based on the same essential structure of rational expectations. Schelling gives a trivial but telling example of how tacit communication can play off the proper meaning, or official communication, in the city:

> ... a motorist at a busy intersection ... knows that a policeman is directing traffic. If the motorist sees, and evidently sees, the policeman's directions and ignores them, he is insubordinate: and the policeman has both an incentive and an obligation to give the man a ticket. If the motorist avoids looking at the policeman, and cannot see the directions, and ignores the directions that he does not see, taking a right of way he does not deserve, he may be considered only stupid by the policeman, who has little incentive and no obligation to give the man a ticket. Alternatively, if it is evident that the driver knew what the instructions were and disobeyed them, it is to the policeman's advantage not to have seen the driver, otherwise he is obliged, for the reputation of the corps, to abandon his pressing business and hail the driver down and give him a ticket. (1960: 149)

The structure of normal communication gives the possibility for all kinds of what Goffman calls 'unratified' communication to take place. In this example tacit signals are involved but more explicit communication consists of hints, innuendos, ambiguities, well-placed pauses and so on, that allow the interaction to carry messages over and above what is routinely going on. These nuances permit new meanings to be conveyed but they are sufficiently close to the structure of expectations of the communication to assure participants of their validity. Unratified communication gives us a good example of how innovations can be made between participants without the surety of established routines of interaction from inherited traditions and social norms.

But where the parties in interaction have no shared history of interaction and know very little about each other, for example, strangers in the city, any unpredictable acts are likely to result in a breakdown of coordination and communication. In this context, the innovative speech or pedestrian act, if surprising to its audience, is likely to misfire and either not be recognized or misrecognised. The act will not 'come off' and will be interpreted as a manifestation of strangeness, which, in turn, prevents the generalization of the moment of creativity across the receiving audience.

This last point shows up the paradox about improvisations and innovations in public (either through speech or action as traditionally understood): the broader the public to which they are performed the less likely they are to come off. Yet the narrower the communities of interest or tradition to which innovation is addressed, the less likely they will be to foster innovative acts because coordination relies on forces such as habit and tradition that are most resistant to change and to the cognitive work required for innovation. Narrower communities are also likely to be the least receptive to innovation because their prevailing norms of interpretation prevent unusual gestures being interpreted as anything other than strangeness or threat.

If this is true there are some salutary lessons for the forms of improvisation and radical street acts encouraged by groups such as the Situationists as ways of awakening the sleepwalking inhabitants of the city and of disrupting the rationalist logic of the Le Corbusian master plan (see Pinder, this volume). Situationist happenings are intentionally radical breaks from the routine structures of communication. As a result the potential community of interpretation for these acts is likely to be narrowly circumscribed. For others, outside the convention of interpretation, their radical meaning might misfire and be interpreted as manifestations of strangeness or threat. In this sense I share de Certeau's skepticism about what he sees as the elitist interventions of such groups, which do not resonate more widely with everyday urban practices.

Reclaiming strategic rationality for the emancipatory city

Instrumental rationality has often been seen as a force of oppression, but this understanding of the dynamics of interaction suggests that it may have an important role to play in the emancipatory city. As we have seen, the norms of interaction among strangers in the city are likely to be secured rationally because there is no recourse to common traditions and customs. But rational coordination is often difficult to settle. Game theorists have shown how hard it can be, even when participants desire it, to coordinate among alternative choices based on rationality alone. Rational assumptions often leave interdependent actors with several mutually exclusive but equally rational courses of action.[1] Without tradition, or at least familiarity, as a guide, it is even more difficult for actors to coordinate. This is one reason why the rather neutral exchanges amongst strangers became the new coordination developed by urban dwellers in the modern metropolis.

The difficulties of rational coordination have dichotomous consequences. Poor coordination can lead to retreat from interaction and communication altogether. In the urban context this might happen either voluntarily, for example, in gated communities, or as a result of wider forces of social exclusion responsible for ghettos. Also the very instrumentalities that lead to indeterminacy are based on self-interest that can undermine more 'social' motives for coordination.

The improvisations that de Certeau points to as an outcome of the indeterminacy of rational coordination can become key anchors in settling new conventions and future expectations. For example, in contemporary western societies there is some indeterminacy around the heterosexual gender norms of courtship and sexual liaison. Who makes the first move? What are reasonable advances to make and in what ways? Are there any clear break points that allow either party to withdraw from the advances? There is, at present, no universally accepted norm. This contrasts with the more settled norms a generation or so ago when courtship, at the local dance hall for instance, was based on a sexist norm – the man always made the first move – but with clear steps in the levels of commitment at which either party could break off the advance without causing mutual embarrassment. Man asks woman to dance; woman normally accepts out of politeness or interest, after which either can break off and not dance together again. Man asks woman to dance a second time; woman can either accept or refuse. If they do dance a second time, the man might then offer to buy woman a drink, and she might accept or refuse. At present, in the heterosexual community rituals of liaison and courtship are more indeterminate, leading to some productive gender role confusions. This might account for the rise in popularity of more structured forms of dating in the city, such as the personal columns in newspapers. These are in some ways the contemporary equivalent of the tea dance, providing a range of possible partners without commitment and a clear structure of meeting that avoids any embarrassment associated with not taking the relationship further: stopping after the first date like stopping after the first dance. In contrast to this, the gay community, who were previously allowed no explicit norms of sexual liaison and courtship, has had to work on them implicitly: at first in secret, from different meeting places in the city to subtle visual cues, and now more openly (see Brown, this volume). As a result, the different sections of the gay community have arguably established conventions of liaison to better effect than the heterosexual community.

Innovative acts are more likely to be successful if they work within existing conventions, or if they provide another way of coordinating activity (see the discussion of focal points, below) when rational assumptions have run out. If new ways of acting in the city (improvisations) can lead to new ways of being in the city more generally, then those new ways of being will have to be norm-bound, that is coordinated rational expectations. Otherwise the hard-won work of emancipatory action will be lost as coordination breaks down and acts are misconstrued.

So we have a conundrum. To make improvisations and stylistic gestures 'stick' as actions, in speech or urban space, we have to have an audience that can receive the message within a convention that gives it force in communication. Otherwise it will be seen as an act of strangeness or a frippery with no emancipatory purchase. However speech acts are more likely to stick in smaller communities of rational expectation. This is why Nancy Fraser (1990) sees focused communities of engagement as the way for disempowered groups to forge effective subaltern counter publics (see also DeFilippis and North, this volume). All this suggests the significance of the politics of scale and the scaling of political activity, which is a central element of the burgeoning interest among geographers in questions of the production of scale (Delaney and Leitner, 1997; Smith, 1992).

The small-scale solution is perhaps the only credible one: not an emancipatory city but a series of emancipatory cities, where freedoms are won within communities of difference. And this is possible because of the overriding rationalist logic of the 'concept city' in which the city is set up as a singularity to serve certain powerful interests. De Certeau (1984) goes further in saying that the singularity of instrumental rationality deploys various 'strategies' with discrete temporal horizons and spatial locations from which they are deployed. He contrasts these strategies of oppression with the sort of more mobile and less durable 'tactics' used in everyday life to resist them. Emancipatory innovation is likely to come from these tactics of everyday life but these tactics, according to de Certeau, are incomprehensible in the face of instrumentalized rationality. By contrast, my argument here is that if the tactics displayed in walking the city are indeed like speech acts then they must have at least some rational support to have force, to 'act', however circumscribed the community of assent may be.

Generalizing the reception for emancipatory acts

I would argue that emancipatory interests are served by making tactics of opposition to the singular instrumentalized logic of the concept city more like strategies, that is, by making them more comprehensible (and hence comprehensive) across communities of difference.

The translation of meaning across difference has been a major theme of political philosophy. I want to take two examples here that work off the two senses of innovation alluded to by de Certeau in the quotation that began this chapter: speech acts in language and pedestrian tactics in the city. Our guide to speech acts in language will be Jürgen Habermas and to stylistic acts in the city Richard Sennett. Their works have very different consequences for understanding the role of rationality in the emancipatory city.

As a legatee of the Frankfurt School, Habermas (1984; 1987) seeks to redeem modern life from the all-pervasive influence of instrumental rationality by reclaiming a substantive notion of 'value rationality' that Max Weber (1968) had argued was lost with the disenchantment of modernity. To do this Habermas returns to the elemental function of language: communication. The notion of the speech act is critical to this project because it provides a basis for distinguishing the different functions of language. There are functions of language on which the criterion of success is the communication of meaning (illocution and locution). But language can also serve as a kind of action in getting people to do things (perlocution). In modern capitalist society the perlocutory functions of language are largely instrumental for individual self-interests. But in speech act terms perlocution is parasitic on locution and illocution (Austin, 1962; Searle, 1979). The way to emancipate speakers from the oppression of instrumentalized perlocution is to create a situation where the purely communicative aspects of language can flourish by ruling perlocution out of court. This is what Habermas calls the 'ideal speech situation'.

Shed of perlocution communication would operate purely as an exchange of meaning. The encounter in public would have to be much deeper than the instrumental exchanges that Simmel regarded as symptomatic of communication in the modern metropolis. Habermas looks to a generalized revelation by speakers of their objective interests, subjective feelings, and social norms so that participants can understand the situated concerns and interests of others and on that basis negotiate new norms common to all. This is where he re-introduces the notion of substantive rationality. In such a fuller public realm speakers make statements reflexively knowing that their validity may be challenged and must be defended on the basis of good reasons or grounds acceptable to others. This 'communicative rationality' contrasts with instrumental rationality that is unreflexive: means-ends based and not open to this kind of scrutiny. Habermas hopes that the unfolding of reflexive discourse will be able to identify common emancipatory and public interests beyond the situations and self-interest of each participant in the conversation.

Habermas identifies here a particular type of emancipatory city with its roots in the cities of ancient Greece. The ideal speech situation mimics the privileged environment of the agora. There, the citizenry – men of property free from menial tasks that were performed by women and slaves, who, in turn, were excluded from democratic participation in the political city – met to decide the issues of the day through rational and democratic debate. Habermas sees a similar space for debate, albeit in a less articulated form, in the coffee houses of eighteenth-century Paris and London where issues of the day were discussed over the new mass circulation newspapers according to rational rules of public discourse.

Richard Sennett (1974) also looks to space in the city to foster emancipatory interests, but he discounts the role of rationality. Using the same examples of newspapers and coffee houses, Sennett points not to rational rules of public discourse, but to the performative aspects of communication in these places. Speakers from different social classes adopted the conventions of the theatre in their verbal expressions and body presentations. Rhetoric and the dramatization of difference were made possible by the adoption of common, artificial modes of speech and action. Sennett points to the importance of an urban public for expressive acts. The urban public realm affords a 'credible space' for different styles of performance.

Returning to the example of speech acts, we can see that for a speech act to come off it is not just a question of style and performance. It must also conform to a convention. As I have argued earlier in the case of interaction between strangers in the modern city, these conventions are rationally supported. They involve the convergence of rational expectations. The eighteenth-century coffee houses were not just home to new ways of performing in public; they were also the home to a new type of public with new expectations about the conventions of interaction in public, at least in those credible spaces. This last point is significant. It was only in certain spaces of the city that the rational expectations could conform to a convention that said dramatic forms of self-expression were permissible and indeed credible.

So again we are left with a paradox. Fuller forms of communication are most possible either where the number and type of participants are circum-scribed (Habermas) or the spaces in which performances can be made are restricted and specialized (Sennett). Whatever the approach, the chances of achieving an emancipatory *city*, rather than *cities*, seem slim. There is also a similarity between Habermas and Sennett in that their visions of emancipa-tion both seem to require a *settled city*: one where discourse and performance can take place under the aegis of substantive rules for debate or credible spaces for performance.

It is at this point in our search for the emancipatory city, that we need to return again to de Certeau (1984), with an added dash of the early Sennett. In *The Uses of Disorder* Sennett (1970) identified the importance of social heterogeneity in the city. The unpredictability of encounter in the city pre-vented city dwellers from settling down, as in some suburbs, into the assumed safety of homogenous communities where fear and prejudice of others could well be fostered. De Certeau also looked to stylistic acts, unpredictable juxta-positions, and disorderly tactics that would and could not be fixed to any stable time or space. Urban dwellers 'act like the "renters" who know how to insinuate their countless differences on the dominant text' (de Certeau, 1984: xxii). These unpredictable, even anarchic, acts in various shifting spaces speak much more to a fluid or *unsettled city*. It is a city of borderlands and shifting spaces, of mutable and uncertain expectations.

What I am proposing here is that it is in the spaces of encounter in the unsettled city that strategic rationality is at its height. Being unsettled is an outcome of the inability to coordinate action based on prevailing expectations (the problem of indeterminacy with instrumental assumptions that we have discussed). In this case hitherto unregarded elements of the environment or discourse can provide 'focal points' on which to coordinate future strategic expectations (Schelling 1960). James Johnson (1993) argues that elements of language can fulfill this role, especially neologisms. Similarly, conspicuous acts of pedestrian style might be the equivalent of linguistic turns of phrase that provide new ways to be, and be understood, in the city. De Certeau gives the example of graffiti:

> the fleeting images, yellowish-green and metallic blue calligraphies that howl without raising their voices and emblazon themselves on the subterranean passages of the city, 'embroideries' composed of letters and numbers, perfect gestures of violence painted with a pistol, Shivas made of written characters, dancing graphics whose fleeting apparitions are accompanied by the rumble of subway trains: New York graffiti'. (1984: 102)

The stylistic 'signature' of the graffiti king fits into a wider vocabulary ('use') among graffiti artists, which in turn disrupts the reading of the city by the wider public.

To allow these plays on official communication to get established as new forms of coordination they need to be generalized. As we have seen Habermas looks to a safe and settled city for deep revelations of the self to occur so new, reflexive understandings can be agreed upon. Likewise Sennett (1974) points to quite circumscribed, credible spaces for new forms of dramaturgical communication across difference to emerge. But when the environment is less stable or certain, communicative action must also be less demanding in its requirements. De Certeau and to some extent the early Sennett (1970) are proposing a more fluid and inexacting framework that recognizes the importance of the surprise gesture, unpredictable encounter, and phatic aspects of speech in initiating, maintaining, and interrupting contact and interaction.

It is here perhaps, where action and communication are most pared down, that we can begin to imagine a way of coordinating action beyond specialized audiences for discrete speech acts. In the unsettled and unpredictable city, in poorly oriented labyrinths of connections and disconnections, it is not just portentous communication that is at stake. The very rules of engagement – the way that contact is initiated, maintained, and closed off – are significant. The way that strangers engage with each other, the degree of openness to the regard of others, the tacit signals that are given off without a word being said – all the simple traffic rules of language could be the place to improvise and experiment, without the pressure of more onerous communication. This first step towards deeper communication is significant because, as this chapter has sought to establish, the structure of communication for implicit and explicit acts is essentially the same. What this approach recognizes is that there is a good deal of work to be done before participants in a Habermasian discourse are constituted as full participants. Equally, not all urban dwellers will feel comfortable with the performative and expressive requirements of Sennett's credible spaces.

The place for experimenting with new tactics for fostering a broader sense of the city and its possibilities is at these initiating edges of communication, where the tone for communication is set. Here innovative speech acts or pedestrian rhetoric might be more effective because the rational requirements of the receiving audience's convention of understanding need not be so onerous because the content of the communication is less portentous. In fact the rational expectation here is simply that there will be different styles of initiation and closure, over body language and small-scale illocutions. And this is where (contra de Certeau) the idea of the whole city might return. The overall convention is that phatic acts are more diverse in the city. Rational action here does not need to judge actions against beliefs but merely the styles of being in the city.

Phatic acts suggest a link between ideas of public space and political 'space' – the public realm. The rational expectation that such rudimentary communications are forms of display rather than significant disclosures is the first step of familiarization between unlike others. It could form the first move in their mutual constitution as fully political subjects. This 'speech place' is based on the ability to be mobile in the range of phatic contacts under a rational convention of styles of tacit engagement. It is why the idea of cosmopolitanism might be a legitimate radical aim. It is also why ideas of public space, involving seemingly superficial contacts, are so important to any final constitution of a political public realm: superficial engagements in these urban spaces are the first tentative steps in speech and street communication towards any fuller citizenship in an emancipatory city.

Note

1 A simple example of this is 'the dating game', which I adapt here from the original 'Battle of the Sexes' example by Luce and Raiffa (1957). Two people have just started dating. The first partner prefers the wine bar on one side of the city, the second partner prefers the pub on the other side. It is more important to both of them that they are together than either are at their preferred drinking haunt. They have both left work but one partner has forgotten their mobile phone. Which bar does each partner head for? There are two rational choice solutions. Since each partner would mostly prefer to be together they either both head for the pub or both head for the wine bar. Both these courses of action are strategically rational in that either partner would not wish to change their choice if they could assume their partner was choosing the same location – whereas if one or both thought they were choosing different locations then one or both would wish to change their choice. The problem is there is no rational way to choose between the two

rational courses of action. In heterosexual partnerships, traditionally, nonrational sexist norms have guided the decision.

References

Austin, J. L. (1962) *How To Do Things With Words*. Oxford: Clarendon Press.

Bridge, G. (1997) 'Towards a situated universalism: on strategic rationality and "local theory"', *Environment and Planning D: Society and Space*, 15: 633–9.

Bridge, G. (2000) 'Rationality, ethics and space: on situated universalism and the self-interested acknowledgement of "difference"', *Environment and Planning D: Society and Space*, 18: 519–35.

de Certeau, M. (1984) *The Practice of Everyday Life*. Berkeley, CA: University of California Press.

Delaney, D. and Leitner, H. (1997) 'The political construction of scale', *Political Geography*, 16: 93–97.

Foucault, M. (1977) *Discipline and Punish: The Birth of the Prison*, trans. A. Sheridan. London: Allen Lane.

Foucault, M. (1986) 'What is enlightenment?', in P. Rabinow (ed.), *The Foucault Reader*. Harmondsworth: Penguin. pp. 32–49.

Fraser, N. (1990) 'Rethinking the public sphere: a contribution to the critique of actually existing democracy', *Social Text*, 25: 56–80.

Goffman, E. (1959) *The Presentation of Self in Everyday Life*. Harmondsworth: Penguin.

Goffman, E. (1983) 'Felicity's condition', *American Journal of Sociology*, 89: 1–53.

Habermas, J. (1984) *The Theory of Communicative Rationality, Volume 1: Reason and the Rationalisation of Society*, trans. T. McCarthy. London: Heinemann.

Habermas, J. (1987) *The Theory of Communicative Action, Volume 2: Lifeworld and System: A Critique of Functionalist Reason*, trans. T. McCarthy. Cambridge: Polity.

Harvey, D. (2000) *Spaces of Hope*. Edinburgh: Edinburgh University Press.

Horkheimer, M. and Adorno, T. (1986) *Dialectic of Enlightenment*. London: Verso.

Jameson, F. (1984) 'Postmodernism or the cultural logic of late capitalism', *New Left Review*, 146: 53–92.

Johnson, J. (1993) 'Is talk really cheap? Promoting conversation between critical theory and rational choice', *American Political Science Review*, 87: 74–86.

Luce, R. D. and Raiffa, H. (1957) *Games and Decisions: Introduction and Critical Survey*. New York: Dover.

Lynch, K. (1960) *The Image of the City*. Cambridge, MA: MIT Press.

Schelling, T. (1960) *The Strategy of Conflict*. Cambridge, MA: Harvard University Press.

Searle, J. (1979) *Expression and Meaning*. Cambridge: Cambridge University Press.

Sennett, R. (1970) *The Uses of Disorder*. Harmondsworth: Penguin.

Sennett, R. (1974) *The Fall of Public Man*. New York: Norton.

Simmel, G. (1995) 'The metropolis and mental life', in P. Kasinitz (ed.), *Metropolis: Centre and Symbol of Our Times*. Basingstoke: Macmillan. pp. 30–45.

Smith, N. (1992) 'Geography, difference and the politics of scale' in J. Doherty, E. Graham, and M. Malek (eds), *Postmodernism in the Social Sciences*. London: Macmillan. pp. 57–79.

Sugden, R. (1989) 'Spontaneous order', *Journal of Economic Perspectives*, 3: 85–97.

Weber, M. (1968) *Economy and Society*, 3 vols. Totowa, NJ: Bedminster Press.

9 Urban Escapades: play in Melbourne's public spaces

Quentin Stevens

It is difficult to generalize about the social life of the city. Urban behaviour does not conform to a simple set of rules, nor does it always meet expectations. The concentration of a diversity of unfamiliar people, objects, meanings, and opportunities for action in urban public spaces stimulates a wide range of behaviours:

> As a place of encounters, focus of communication and information, the *urban* becomes what it always was: place of desire, permanent disequilibrium, seat of the dissolution of normalities and constraints, the moment of play and of the unpredictable. (Lefebvre, 1996: 129)

In this chapter I consider a variety of cases where urban settings frame playful behaviour. I explore the concept of play, as an overarching framework for understanding different kinds of freedoms that exist within social practices in urban spaces.

Play is an escape from the pursuit of instrumental purpose. It is pleasurable in itself and not because of any particular outcome (Bataille, 1985; Gilloch, 1996: 84). Play is superfluous activity, and as such people are free to choose their level of involvement. Social relations in play remain open to negotiation and change. Play is predicated on a degree of freedom from obligation and coercion. Play often occurs within physical boundaries and follows special rules that define 'a place apart'. This separation allows people to escape their everyday roles, conventions, demands, and restrictions (Goffman, 1972; Huizinga, 1970). Especially in urban spaces, play often occurs in the presence of a diversity of strangers, taking one beyond habitual patterns of social engagement.

Drawing upon Thrift's (1997) account of non-representational theory, I will suggest how playful practice is a 'performative experiment', where the joint actions of individuals, shaped by the dynamic contingencies of their spatial encounters, actively define social potential. The expressive, affective and perceptual powers of the body make possible the enactment of different ways of being in the world. I argue that the processes of discovery, invention, and transformation inherent in play are concrete realizations of freedom.

Play behaviour can be divided into four distinct types: competition, simulation, chance, and vertigo (Caillois, 1961). Each type of play highlights different ways in which play provides an escape from the conventions of social life. Through each of these forms of conduct, the subject is configured in particular relations to their own body, to space and to other people. Competitive play involves tests of strength and skill. It is an escape from instrumentality, because it commits bodily effort to non-productive purposes. In competition, people forget their everyday concerns and responsibilities, and explore a particular aspect of self. Conflict becomes an opportunity for transformation, by pushing the limits of what the body can do.

Simulation involves disguising or forgetting one's usual self and one's place in the world by fabricating other identities and situations. Such imaginative performances allow escape from social and environmental limitations. Meanings constituted through such interpretive acts remain ephemeral and ambiguous, and are hence difficult to regulate.

Play as chance means abandonment to uncontrollable and unpredictable circumstances. Spontaneity and novelty provide opportunities for escape from predetermined and ritualised courses of action. Vertigo includes acts through which people escape normal bodily experience, 'losing themselves' in a purely physical mode of being, free of social meaning and purpose. Vertigo is found in settings and forms of activity that destabilise or alter perception. Such experiences confound clear, logical apprehension. Through psychological forms of vertigo – crude, disorderly and destructive behaviour – people turn against social propriety, acting out a release 'from the terrors of social responsibilities and pressures' (Caillois, 1961: 51).

These four fundamental types of play are different ways in which life is lived more intensely. Competition and simulation focus on increased personal control over the body and over communicated meaning. Chance and vertigo involve escape from behavioural and perceptual controls. The routines, constraints, and preconceptions of everyday social reality are necessarily interpreted as they are perceived and performed by human bodies in the context of specific spatio-temporal encounters. Each of the four kinds of play employs the potential of the body as a sensitive and expressive mechanism which explores and effectuates the degree of freedom or 'play' within these rules. Freedom is practised as people test the limits of experience in these different ways. I will examine in turn examples of each of the four kinds of play.

Competition

Two men play chess on a giant board set into the pavement of a small plaza on one of Melbourne's main pedestrian promenades (Figure 9.1). By focusing all their attention on a test of mental skill, the men suspend the limits of social identity, as defined by such traits as race, class, and occupation. Within the

FIGURE 9.1   A public chess game (Photo: Quentin Stevens)

chess game, the participants have freedom to direct their own action. Submission to the rules of the game is an essential part of this escape. However, these rules are not submitted to the logic of efficiency and accumulation. Such kinds of competitive play behaviour loosen the idea of function: they demonstrate effort without material gain. This public setting frames an intellectual escape.

Because play in this public context involves an often unknown and unpredictable opponent, the game is an exploration. It structures an escape from conventional social engagements that allows the players to encounter each other. Being on public display heightens the tension of this encounter, pushing the players to excel.

The chess game also frames an opportunity for other people to forget about their everyday concerns. In the wider context of the plaza, the freedom of play is captured for profit: the game becomes a spectacle consumed by patrons of the overlooking café. However, passing pedestrians are also distracted from their instrumental concerns. They stop to watch; some sit on the steps. They comment to their companions or even step forward to offer advice to the players. People exercise the freedom which a public setting provides to raise their level of involvement in the contest, and to move between the roles of audience and participant.

The freedoms of play are not always neatly regulated by rules of conduct or safely circumscribed within a place apart. Another example of competition shows how playful practices can generate tensions with other more instrumental uses of urban public space.

During rush hour on Friday evening, hundreds of cyclists ride together through the centre of the city. Each month this 'Critical Mass' follows a new route, unknown even to most of the participants. They gradually explore the city. Their growing understanding of urban space is essential for freedom in using it. They occupy the whole street and ignore traffic signals, suddenly and temporarily blocking vehicular traffic at a series of major intersections. This event is a political critique, promoting the right of cyclists to free use of the streets through direct confrontation with car drivers. They resist the road laws which are imposed on this urban space. As the movement's name suggests, the sheer numbers of cyclists in Critical Mass provides a certain power to exercise their claim for space in the face of rush-hour traffic.

While car drivers are hindered in their efforts to get home and relax for the weekend, the cyclists are relaxed and enjoying themselves right there in the city streets. They roll along slowly, chatting to each other, ringing their bells; they're not really aiming to get anywhere. As a playful escape from efficient transport, Critical Mass is a critique of the idea that human movement should be functional (Lefebvre, 1991a; 1991b; 1996).

FIGURE 9.2   A collective bike lift during Critical Mass (Photo: Quentin Stevens)

A collective 'bike lift' at a major central city intersection (Figure 9.2) highlights that Critical Mass is not serious cycling. This action is playful in a number of respects. It parodies the practicality of transport. It is an inefficient engagement between man and machine, and a bodily challenge. The inversion of bicycle and rider transfers attention away from the contentious vehicle and onto the person who controls it. This emphasizes the commonality between the actors and their audience of onlookers. The bike lift is also an unexpected use of place. This playful performance also serves an instrumental purpose as a contesting of space. Lifting the bikes in unison is a metaphorical show of their collective strength. The political symbolism of this act is amplified by its location directly in front of the Town Hall.

The cyclists rely on a broader tolerance of diverse social behaviour in an urban setting. The guarantee of Critical Mass's freedom to exhibit their politics is only tacit, expressed in the smiles of watching pedestrians and the forbearance of gridlocked drivers. By exercising the scope of social rights in the use of public space, these cyclists affirm and indeed expand them.

There are many tensions in Critical Mass, an act which combines leisurely, unstructured exploration with a determined, instrumental effort to hinder the needs of everyday car commuters. The cyclists' play is not just an escape from the social regulation of a space; it temporarily alters the structure of use of that space. Behaviour through which the cyclists find freedom occurs at others' expense.

But giving priority to the cyclists' transgressions and reversals, such as the bike lift, cannot be justified on rational grounds, because the cyclists' actions go beyond conventional understandings of rational need and action. Their emancipation comes both through oppositional struggle and through the fluidity, multiplicity and spontaneity of their new practices (Jackson, 1988). While a range of inventive, playful, bodily actions are used to defuse the atmosphere of conflict, they are also a most effective means of engaging in conflict. The forms this encounter takes shows cyclists getting a great deal of bodily enjoyment out of confrontation itself. In this competition, social relations of play are intertwined with relations of power. Play itself takes part in the struggle over the freedoms available in urban society. The definition of appropriate behaviour in urban social space is constantly being reshaped and reinscribed by such playful practices.

**

While one of the examples of playful competition I have discussed followed strict rules, the other re-wrote social rules. Both cases occurred in public settings, where people were exposed to others with different abilities and interests, including audiences who experienced such contests from a distance. These escapist play acts were prompted and shaped by the social diversity gathered together in urban public spaces, and the tensions which arise in the uses of those spaces.

Simulation

The public artwork *The Three Businessmen Who Brought Their Own Lunch* (see Figure 9.3) is located at Melbourne's busiest pedestrian intersection.

FIGURE 9.3 Sculpture: 'The Three Businessmen Who Brought Their Own Lunch'. The rear figure has a lit cigarette in its mouth (Photo: Quentin Stevens)

It consists of three very thin, life size bronze statues wearing suits and carrying briefcases. They stand on the kerb and stare quizzically out at the city.

People invent many different ways of playing with the statues, all of which pretend they are real people. They explore the fine details of the sculptures' physiques through both vision and touch, and discover what the statues can 'do'. People stand arm-in-arm with the figures, hug them, imitate their stiffness and their comical facial expressions. They shake their hands, pick their noses and pat them on the belly. Many of these playful engagements are transgressive of behavioral norms. They are performances of imagined social relations with strangers which are inappropriate within urban society, particularly as these are businessmen.

People's playful contributions also imagine new roles for the statues. In winter one is given a woollen hat. All three of the figures have been designed with mouths pursed into deep circular holes. A woman leaves her lit cigarette in the mouth of the third statue, and she and a friend have a laugh, recognizing that passing strangers are confronted by her contribution. Public statues usually express society's higher ideals. Adding a cigarette transforms these sculptures into a promotion of something profane.

Many of the tensions of urban life are written into the Businessmen. They appear tense, harried, expectant. Their formal dress and posture contrasts with the humour of their exaggerated features. The looks of surprise and apprehension on their faces, their frail bodies and unsteady, tilted stance suggest an inadequacy. Their identity is completed and extended through interactions with real people. These sculptures provide an opportunity for people to make their mark on the lives of an imaginary other. It frames a pro-active role within a close encounter, in the undemanding mode of let's pretend.

In some cases, play involving the statues happens without an audience. This suggests that people engage in simulative play not just as a display to others, but to test their own bodily skills, as an escape into fantasy, and even just for its own sake, for the pleasure of the bodily experience. This artwork makes possible escape from everyday behaviour because, like most public art, it lacks 'function' in the strict sense; it doesn't help achieve any specific practical outcome. The representational purpose of these three figures is unclear.

These figures can easily be drawn into simulative play because they are carefully scaled within an urban theatre. The statues are on the footpath at eye level, close to where people stand waiting for trams. They can easily be reached and reached around. People are free to inspect them up close, to touch them, and to treat them irreverently.

Such artworks in urban spaces inspire unanticipated activities. Even if only for a brief minute, the figures totally absorb people's attention, diverting them from their instrumental responsibilities and concerns.

Just as civic design can encourage people to expand upon the meanings of bodily practice, ritual public practices can lend meanings to spaces (Lefebvre, 1991a; 1991b). Formal parades along Melbourne's main street axis, Swanston Walk, pass by a number of major civic institutions, drawing them together into narratives which bind social identity to place. Melbourne holds a street parade as part of the annual festival called Moomba. Traditionally, this was 'the people's festival', centring on the active participation of a large number of community groups. Hundreds of costumed people marched or danced along, accompanying thematic, musical floats which had been decorated by the groups themselves. The parade gave licence to a wealth of bodily experiences, and displayed and invigorated the city's ethnic and social diversity (Brown-May, 1998).

In 2000, this parade was replaced by a procession of decorated trams containing professional performers. The themes of the trams were playful re-interpretations of aspects of local urban culture, and evocations of the abandonment and social transgressions and inversions of carnival: yet another case where 'what was once intensely lived becomes mere representation' (Debord, 1994: 12). This simulation of freedom masked the production of behavioural controls. The public became marginalized in the role of passive spectators. Public leisure was carefully choreographed, on the very day which is meant to sanctify the idea of public 'free time': the Labour Day holiday.

However, even in this context, the spatial action of certain individuals succeeded in freely expressing different social meanings. At the same time as the trams passed down the middle of Swanston Walk, someone moved along the footpath dressed as a giant budgerigar. It seemed to be part of the

Moomba parade. But the people accompanying it were chanting, 'Bring back the Bird!' Later the same day a person wearing this costume jumped from the prominent bridge where the parade route crosses the Yarra River.

These protesters sought to draw attention to the cancellation of another popular event which had for many years formed a significant part of the Moomba festivities: the Birdman Rally. In this competition, people launched themselves in home-made, unpowered aircraft off the side of a city bridge and attempted to pilot them over a set horizontal distance. For many years everyone crashed short. Many participants took off in nothing more than a funny costume, wildly flapping their arms in a ludicrous imitation of flying. The ungainly bird maquette ably represents the whimsical spirit of this contest. The Birdman Rally was a grand example of public play. It was a participatory event which brought together competitive display, intense, risky experience of the body in space, and sudden, dramatic wasting of energy. Although broadly regulated, it promoted freedom, the pushing of the limits of human experience. The budgerigar, a common household pet, is an ideal rallying symbol of this quest for free flight.

The Birdman protesters struggle against the curtailing of behavioural excess at Moomba. They do so by harnessing the social setting of the formalized procession and turning it against itself. They appropriate the audience gathered along the path of the main parade by running their own event parallel to it. Their march simultaneously challenges the main parade and attempts to look like a part of it, through the use of a giant, fun, colourful figure. Through playful performance, the symbolic terrain of Swanston Walk becomes contested, freeing up the physical and representational potential gathered on the axis.

This performance also offers a critique of the organized parade as a passive consumption experience. Moving along the footpath, the group are closely engaged with the crowd. They invite people to join them. Their display is unanticipated and their intention obscure. They bodily re-invent the parade as an active celebration of nonsense, imitation, and vertigo.

This group's informal parade illuminates tensions which lie at the heart of street carnivals, and public play in general. Carnival is a symbolic rejection of social norms of prudence and propriety. It relieves discontent by providing for the release of pent-up energies and indulgence in forbidden desires:

> carnival celebrated temporary liberation from the prevailing truth and from the established order; it marked the suspension of all hierarchical rank, privileges, norms and prohibitions ... a special form of free and familiar contact reigned (Bakhtin, 1984: 10).

By channelling society's excesses and contradictions into discrete times and harmless forms, the escapist rituals of carnival help to enhance stability, and thus preserve the wider social order. Yet they are not reducible to this order

(Jackson, 1988; Lefebvre, 1991a). The actions of the protesters reveal the difficulty of maintaining social cohesion through carnival. Symbolic forms of freedom and licentiousness inevitably stimulate the body, arousing desires and new, impetuous, practices. The protest also illustrates how an experience of freedom can take shape oppositionally, in the transgression of imposed meanings and behaviours. These people don't take Moomba's fun seriously. They dispute one playful simulation with another. The Birdman protest co-opts the playful imagery, atmosphere and actions of the Moomba parade, but subverts its purpose. The protest's realignment of the playful and the serious tips the balance of this public carnival away from validation of the established order and toward its rejection. Their protest helps other people escape from the structured spectacle, where their only role is as passive consumers. It asserts people's freedom to decide what should be celebrated in public and how.

The ritual practice of parading along an axis serves to regulate people and meanings in space. In the earlier example of the chess players, the spatial ordering of the gameboard itself provided a freedom to pursue playful social encounters. The budgerigar and protesters located themselves on the same chess board, for the reason that it is well sited for a public display. In this instance, the enactment of their freedom from social convention completely disregarded the rules of the chess game. Simulative play occurs in a world apart from obligatory definitions. Through such play, people freely draw upon the diversity of meanings that pervade urban public spaces. People engage with spatial meanings that suit their fancy; they ignore some representational contexts and turn against others. Meanings certainly adhere to the built environment, but they cannot be fixed there and cannot be enforced. The acts of simulative play I have described reveal the inalienable freedom people have to take advantage of opportunities to explore and redefine meaning, to construct new and fantastic perspectives on the world.

Both the examples of simulative play that I have described show how urban public spaces encourage escapes from present circumstances through both memory and fantasy. The city provides a myriad of settings and objects which catalyse this imaginative form of play. Yet the built environment is also constantly being freed from existing contexts of meaning, and given new significance, through the ways that it is touched and represented in the playful actions of human bodies.

Both simulative and competitive displays are stimulated by the presence of strangers. Urban public spaces help structure relations of performer and audience, while they also allow people freedom to move between these roles, and to control their level of involvement (Lennard and Lennard, 1984). The sheer diversity of persons and social contacts in the city undermines the constraints of social conventions (Bourdieu, 1977: 233). Performances in front of an unknown public offer a certain degree of freedom to define one's social identity. Yet the meaning and functional potential of the self depends upon its reception by others, who are also able to imagine and evaluate.

Expressions of difference through play don't necessarily breed tolerance or long-term change. Thus competition and simulation illustrate 'a world of permanent dialogue' (Thrift, 1997: 138) between presentations and their affirmation by audiences of strangers.

Chance

A man stands out in the middle of the footpath on one of the city's busiest intersections, holding up a large placard explaining his personal political philosophy. His position indicates that he is keen to attract attention and engage passers-by. It is in such a densely-used public space that he can make contact with the greatest diversity of people to generate friction and stimulate debate. In the city, strangers' paths inevitably cross. Such encounters are unplanned, and this changes the nature of social engagement, making it less predictable, more risky, and potentially more provocative. Through chance encounters with difference in public spaces, people free themselves from the security and conventionality of their everyday social experience.

Every once in a while, someone stops and steps forward to engage the man; he cannot guess their point of view. This element of chance, an escape into an unpredictable, relatively unrestricted social involvement, is part of the thrill of standing out there. What makes chance so captivating is the tension brought on by the risk of the unknown. Unstructured urban encounters frustrate expectations about how other people will behave. Exposure to unfamiliar impressions and events in public spaces opens up potential for the man to improvise his own behaviour, freeing him from habit (Bourdieu, 1977). The negotiation of these encounters make demands upon the man's intellectual resources and social skills, stimulating new performances of self which engender his development as a social being.

Engagement with this unusual individual can also be enjoyable for passers-by. He distracts them from their premeditated, instrumental courses of action, and makes them temporarily forget their everyday concerns. Many glance up and read his statement. It's hard to know what to make of it; it challenges their frame of mind. His sign presents an invitation to open an encounter (Goffman, 1972), but it is a form of encounter to which all parties are somewhat unaccustomed. The responses of those who engage the man are spontaneous: they elude preformed opinions, require free thinking, and engender an unfolding discourse.

Although the man's action is confrontational, the situation is playful because passers-by have a certain amount of control over whether they engage with him. These people approach by choice. In doing so they stimulate and explore their own curiosities and abilities. They can determine how involved they get, and are free to leave when the discussion no longer gives them pleasure. These interactions are exploratory and don't serve any long-term goal. They are escapes from necessity. A few people choose to linger, even though it's a work day, and indulge in discussion in the midst of a flow of strangers. People who engage him generally stay and talk even after the crossing signals have run a full cycle.

The majority of pedestrians keep a wary distance from the man, and keep to themselves. Chance remains playful when people can understand and choose to accept the extent of risks. However, the consequences of playful public encounters can never be submitted to a rational calculus, as they are a joint outcome of the resources, desires, and actions of various separate parties. Anonymity breeds uncertainty, and in this case, potential for spontaneous expressions of opinion is stifled by the imagined risks of unstructured interactions.

However, this particular urban setting works against perfect isolation from chance incidents. The tight, busy space of the intersection means people's engagements with the man and his sign are often close and sudden. Such close encounters arise spontaneously, as people find themselves brought into close proximity with a stranger when they are stopped at the light or moving through on the cross-street. These engagements are unexpected. They are therefore generally without instrumental intent, and are thus more likely to be exploratory and playful. Indeed, these engagements occur contrary to people's general intentions to keep a civil distance from strangers. This kind of proximity leads to intense visual, auditory, olfactory and even kinesthetic encounters, which transgress the boundaries of public decorum. In this way, urban density frames the possibility of distraction, of unfocussed, often involuntary bodily perceptions of others, which elude rational appreciation and control, and free people from the limitations of their own propriety.

Chance is a fundamental part of the urban condition. It is closely linked to the city's diversity, intensity and irrepressible dynamism. Public spaces do not force order upon social life. Intentions, functions and rules overlap and contradict one another within public spaces; they are not subordinated to any overall rationality. Expectations are often disturbed by urban experience. Urban life offers both spontaneity and deviance. The strangers who surround us have different abilities and objectives. Their freedom of action can lead to unanticipated and unpredictable social encounters. The complex interlacing of public pathways and the fine-grained mixing of activities in the city frame many possibilities for chance exposure to strangers and to new experiences. The structure of the city thus contributes to the unstructured nature of city life. This indeterminacy is a measure of freedom from premeditated behaviour. Action evolves on the spot, shaped by the contingencies of people brought together in time and place.

Unplanned, uncontrolled experiences in urban space defy rational appraisal and response; they provide scope for the stimulation and expression of repressed desires (Savage, 1995). Chance experiences thus frame freedom from 'the unquestioned virtues of sobriety, industry, rationality, diligence and so forth' (Lofland, 1998: 121).

Vertigo

On a summer Sunday, loud pop music spills from a store on the pedestrianized Bourke Street Mall. It saturates public space and distracts people's attention. The music attempts to awaken desires in the body, but only in order to capture them for private gain, by channelling them into consumption. However, commercial interests cannot dictate what actions result. As the music escapes from the threshold, it is unleashed from specific forms of production or consumption.

A man in his fifties dances exuberantly in front of the store. He is quite athletic and quite uninhibited. The man's vertigo expresses the freedom people have in their bodily engagements with the many sensory stimuli which are compressed together in urban space, which includes not just sounds but a wealth of images, smells and textures. This example shows how a playful experience of the body's capacities can be aroused by the body's exposure in the urban environment.

Sound is insubstantial and ephemeral; its pleasure lies not in being possessed, but in being experienced. The man's dancing is a way of coming to terms with the music, which stimulates the body. Sometimes he accompanies the tune on a harmonica. The man's responses suggests that music 'found' in the public realm is not a sacred, finished piece of work, but remains open to interpretation. His playing establishes a reciprocity between the music and his own desires.

The man escapes everyday, practical behaviour by attuning his body to the music's rhythms. By remaining in the public space he retains the freedom to respond how he wishes, freedom to move, to jump and spin. He generally has his back to the store. Rather than yielding his attention to the merchandise, his use of the sound draws attention to him. The man invites passers-by to join him, and several do. His enthusiasm rubs off. This performance of vertigo itself becomes a stimulus, demonstrating a behavioural possibility to others.

Some other forms of vertigo are more risky and more at odds with everyday uses of public space. At Town Hall Plaza, two teenagers on in-line skates get up speed and then jump clear of the ground, grinding the base of their skates along the edge of a large stone step, as well as down the bumpy edge of the adjoining sculpture (see Figure 9.4). These skaters only see the giant public chess board below their feet as a smooth surface; their playful moves across it have no regard for the rules of that game. The whimsical connotations of the sculpture are lost on them, and although they make use of the steps as a place to sit and enjoy the behaviour of others, their creative, exploratory action also lends the steps a new kind of usefulness. The physical environment of the city does not dictate the playful possibilities of skateboarding. Urban space is, rather, a 'drawing board' on which potentials for spatial engagement are perceived and acted out. The freedoms of the city are

FIGURE 9.4   In-line skaters exploring the urban landscape
(Photo: Quentin Stevens)

defined through the projection of the body's own speculative spatial gestures within a complex terrain of boundaries, levels and inclines (Borden, 2001).

The psychological element of vertigo becomes apparent where play acts like skating confront the behavioural norms built into a setting. In an attempt to prevent the disruptive and destructive play of the skaters, the city council subsequently installs projecting metal lugs at regular intervals along the edge of the steps of this plaza, as well as ledges in many other locations throughout the city. But this doesn't stop skaters or make them go away. This restriction engenders a creative response, as skaters invent new kinds of games to test their skill. Skateboarders make faster approaches, leaping for higher, slipperier and more risky edges, such as handrails. They jump out off ledges or over small flights of steps. Skaters who leap into the air seldom land safely, but that's scarcely important. Not all people use steps and handrails for safety. Skateboarders also grind along the smooth top edge of the metal lugs themselves. In one dramatic case, a lug is bent over sideways from repeated heavy impact.

Skaters don't skate because it's easy. They seek out physical challenges in the landscape, and the intricacies of the city offer so many. Skaters have intensely physical engagements, escaping from bodily passivity. Their play practices develop in dialectical relation to the limits the physical space imposes, defining new forms of freedom through the testing of constraints. Somewhat paradoxically, design features which make urban public space safe and functional inspire the most extreme experiences of escape, by opening up possibilities for rapid movement and for intense sensory stimulation.

Risk is the key element that makes such actions an absorbing and fun escape from normal bodily experience. Risk exists because skaters relinquish the slow, steady, predictable connection to the ground. Their behaviour goes beyond the controlled patterns of movement around which urban space is designed. Taking off from a high ledge, skaters experience the thrill of vertigo as their elevation is converted into speed in an explosive release of potential energy. With increased excitement comes increased danger. The vertigo of skating involves exposure to real risks. Freedom also means a lack of security. For such actions to remain enjoyable and playful requires some limit to consequences. Skaters retain a measure of control over their encounters with the uncontrolled. They choose challenges which provide 'a little more risk than can easily be handled' (Goffman, 1972).

The risks attendant to the skater's speed and trajectory and the physics of wheels shifts their focus of attention. Skaters know the urban terrain primarily through the sense of touch (Borden, 2001). They focus on the swift and pre-emptory reading of texture, inclination, and elevation ahead. Such tactile experiences of the city subvert rational appraisal and engagement, by substituting a focus on immediate, fluid movement for any external goal. Public space has an erotic quality; an intensity of unfamiliar phenomena constantly assails bodily senses which do not submit to focused rational consideration and instrumental management, arousing desires which are irrational and often suppressed (Latham, 1999; Lofland, 1998: 121–4).

There is playful escape in this re-reading of a landscape which the average pedestrian takes for granted. Their velocity demands precise coordination of the body. Grinding and jumping are just two ways that skaters push the limits of what can and cannot be done in urban space.

Urban space frames a density of overlapping, multisensory perceptions which can intoxicate the body. Playful acts of vertigo such as dancing and skating heighten the bodily thrills and risks of engagement with the intricacies of the urban physical environment. The design of public open spaces does not neatly govern bodily action in the interest of practical function. Hence the city situates a range of new kinds of engagements between body, self, and space.

The psychological aspects of vertigo are stimulated by the social intensity of urban public spaces where a myriad of social activities overlap and interpenetrate (Sennett, 1971). In public settings, most people remain strangers, maintaining a civil distance and becoming only loosely engaged with each other (Lofland, 1973; 1998). Psychological vertigo becomes a means through which individuals gain agency within this social world. Paradoxically, such unregulated, free practices serve to make strangers more involved with each other, heightening mutual awareness and tension. What psychological vertigo provides escape from is the alienation and indifference of the city.

Conclusion

Melbourne's public realm frames a rich and fluid context of people, social meanings and physical landscapes. My examples of four different types of play illustrate how people's actions, drawing upon particular configurations of the urban context, give substance to the idea of freedom.

In the city's public spaces, people encounter a diversity of physical and intellectual challenges. In meeting such challenges, people behave in ways that elude instrumental conceptions of function. They expand their own abilities through action. Such displays often come into conflict with existing practices. The examples of playful competition I observed embodied new ways of relating to difference.

The city is filled with symbols that define social values. Some of these are unfamiliar to the viewer. They provide opportunities for imaginative interpretation through simulation, which transcends conventional meanings. The coupling of anonymity and visibility in public can frame opportunities to escape from conventional social roles and perform new understandings of self.

The intricate physical settings of the city frame chance exposures to new experiences, which may distract people from their predetermined purposes and provide opportunities for new knowledge. Because they must be improvised, chance social encounters elude familiar forms of social engagement.

The city exposes people to a wealth of new physical sensations. Its fast pace, compactness and complexity lead to excitation of the body, through which people escape from composure. Play as vertigo is both stimulating and dangerous. I showed how people made use of both the rhythms of the city and the complex surfaces of its public spaces to push themselves beyond safe and predictable ways of moving. In the case of skateboarding, part of the freedom of play was its confrontation with dominant notions of acceptable practice.

For each of the four different types of play, people's exposure to strangers in urban spaces makes an important contribution to the potential for creative activity and for escape. Freedom does not necessarily come through isolation. The expressive actions of others provide opportunities to make sense of the world. It is only through the presence of other people that certain forms of behaviour become possible.

Urban space gives structure to social encounters, bringing people together in ways that heighten their experience of other life possibilities. The openness, multiplicity and layering of urban public space can be used to control one's level of exposure, to limit the seriousness of consequences. The anonymity of the individual provides a general freedom of action in public space. Away from the assigned roles and enclosed spaces which can regulate behaviour in the private realm, those who play can escape censorship, if not judgement, by others (Lofland, 1998). They can, by choice, display and develop particular aspects of self and distinctive modes of being in the world.

The example of Critical Mass showed the public as an audience which recognized and validated such performances. In the case of the Birdman protest, the play of strangers also attempts to persuade those who watch to understand the social world in a new way, and hence to transform their own practice, to resist the regulation of escapism.

It is while people are circulating through urban public space that they are most at liberty to take on new roles as audience, or even as actors. The dancer in front of the music store inspired and encouraged others to copy him; this outcome shows people exercising the freedom to move between these kinds of roles. Indeed, people's visibility in public settings, such as the site of the *Three Businessmen* statues, inevitably projects them into the role of performer, breaking down the boundary between seeing and being seen. Exposure can generate responses of self-consciousness and withdrawal, not only freedom and abandon. The openness of behavioural possibility in public is always a test of people's disposition toward novelty, challenge and risk.

Strangers encountered in public also provide points of contrast, which accentuate and stimulate our own demonstrations of self. The differences in other people's behaviour establishes both limits and possibilities of experience. In the case of the public chess game, strangers were antagonists who pushed players to excel in their own capacities. Here, freedom to achieve is defined within a tight framework of rules. The man standing in mid-footpath with his placard was an instance of playful confrontation where social rules had to be made up as the people went along. In public spaces, the freedoms of presence and action which belong to others lead to unexpected and unusual encounters. These playful, non-instrumental engagements with strangers are opportunities for people to confront and change their own inhibitions, biases and beliefs.

A final consideration, reinforced by the observation of skateboarders, is that other people in the city are busy trying to be free. And they are free on their own terms; they don't see or act in the world in ways which conform to norms. They don't necessarily recognize the conditions and definitions which other people's actions, playful or serious, have inscribed into the world. Attempts to control the playful behaviour of others, who use the city differently, can in themselves propagate new conceptions of free practice. The most difficult freedom of the city is that every significant act may be ignored, repudiated or erased by the explorations and actualizations of others at play.

Public spaces and social rituals are always to some degree free from rational administration, and loosely defined within social mythologies. They always remain available for appropriation and incorporation into new and at times confrontational practices. In urban spaces it is difficult to organize and constrain behaviour, because so many possibilities are readily available. It is also difficult to exclude new perceptions, which press in around the body.

Within the intricacies of everyday urban experience, many conflicts and contradictions inevitably arise. Perhaps the most comprehensive way to characterize play is as a set of means of overcoming such difficulties. In practices of play, people give effect to their critique of everyday life in the city (Lefebvre, 1991a). They escape, if only temporarily, its pressures and hardships.

Possibilities of escape through play become more obvious and more dramatic the more closely they are counterposed with seriousness, functionality and control (Bakhtin, 1984). The anonymity, unfamiliarity and intense stimulation of urban life can be experienced as problems which threaten individual freedom and limit action (Simmel, 1997). I suggest that such conditions can also, in the mode of play, become opportunities for playful actions which are expressive, creative, and free. Baudelaire viewed the city as:

> a site of intoxication ... home to the unexpected, to novelty and distraction, ... a space to be explored with joyous abandon. It offers the excitement of the anonymous crowd, the exhilaration of freedom and the ecstasy of losing oneself. It is a place of shock. (cited in Gilloch, 1996: 171–2)

In Baudelaire's formulation, the shocks of the urban frame the possibility of freedom from expectations and everyday social roles, and escape from conventional and rational approaches to life.

Conventional understandings of play itself are not immune to the possibility of disputation through exploratory practice. The diversity of forms of play taking place over the public chess board reveal the limitations and tensions of different people's efforts to escape the routines of their everyday lives. Even a space specially designed around the idea of leisure only serves certain freedoms. Events surrounding the Moomba parade showed that even the physical openness, fantastic symbolism and contrived exuberance of urban leisure settings are turned to new and unexpected purposes through acts of play.

In different cases, playful actions may subvert, loosen, or transform presupposed rules of social conduct. These performances are not without their conflicts and limitations. The freedoms of play are invariably in tension with the organized schema of 'serious' or dominant beliefs, expectations, aspirations and capacities which are defined and reproduced by every bodily practice (Bourdieu, 1977; Foucault, 1997). Huizinga (1970: 24) notes that 'the contrast between play and seriousness proves to be neither conclusive nor fixed'. The circumstances which surround people's playful actions in urban public space are dynamic, and the freedoms which are enabled through them are fleeting, partial, and uncertain. While the elements of my definition of play point toward many possible expressions of emancipation, they can never guarantee them. Freedom can be found in the city not because ideas, resources and constraints are obvious and permanent; in fact, quite the opposite. The incoherence and flux of the city are the strongest guarantees that there will always be some 'play' within the order of things.

References

Bakhtin, M. (1984) *Rabelais and His World*, trans. H. Iswolsky. Bloomington, IN: Indiana University Press.

Bataille, G. (1985) 'The notion of expenditure', in A. Stoekl (ed.), *Georges Bataille: Visions of Excess – Selected Writings, 1927–1939*, trans. A. Stoekl, C. R. Lovitt and D. M. Leslie, Jr. Minneapolis, MN: Minnesota University Press. pp. 116–29.

Borden, I. (2001) *Skateboarding, Space and the City: Architecture and the Body*. Oxford: Berg.

Bourdieu, P. (1977) *Outline of a Theory of Practice*, trans. R. Nice. Cambridge: Cambridge University Press.

Brown-May, A. (1998) *Melbourne Street Life*. Melbourne: Australian Scholarly Press.

Caillois, R. (1961) *Man, Play and Games*. New York: Free Press of Glencoe.

Debord, G. (1994) *The Society of the Spectacle*. New York: Zone Books.

Foucault, M. (1997) 'Space, knowledge and power' (interview conducted with Paul Rainbow), in N. Leach (ed.), *Rethinking Architecture: A Reader in Critical Theory*. London: Routledge. pp. 367–79.

Gilloch, G. (1996) *Myth and Metropolis: Walter Benjamin and the City*. Cambridge: Polity Press.

Goffman, E. (1972) 'Fun in games', in E. Goffman, *Encounters: Two Studies in the Sociology of Interaction*. London: Penguin. pp. 15–72.

Huizinga, J. (1970) *Homo Ludens: A Study of the Play Element in Culture*. London: Temple Smith.

Jackson, P. (1988) 'Street life: the politics of carnival', *Environment and Planning D: Society and Space*, 6: 213–27.

Latham, A. (1999) 'The power of distraction: distraction, tactility and habit in the work of Walter Benjamin', *Environment and Planning D: Society and Space*, 17: 451–73.

Lefebvre, H. (1991a) *Critique of Everyday Life, Vol. 1*, 2nd. edn., trans. J. Moore. London: Verso.

Lefebvre, H. (1991b) *The Production of Space*, trans. D. Nicholson-Smith. Oxford: Blackwell.

Lefebvre, H. (1996) *Writings on Cities*, trans. and eds E. Kofman and E. Lebas. Oxford: Blackwell.

Lennard, S. and Lennard, H. (1984) *Public Life in Urban Places*. Southhampton, NY: Gondolier.

Lofland, L. (1973) *A World of Strangers: Order and Action in Urban Public Space*. New York: Basic Books.

Lofland, L. (1998) *The Public Realm*. Hawthorne, NY: Aldine de Gruyter.

Savage, M. (1995) 'Walter Benjamin's urban thought', *Environment and Planning D: Society and Space*, 13: 201–16.

Sennett, R. (1971) *The Uses of Disorder: Personal Identity and City Life*. Harmondsworth: Penguin.

Simmel, G. (1997) 'The metropolis and mental life', in N. Leach (ed.), *Rethinking Architecture: A Reader in Cultural Theory.* London: Routledge. pp. 69–79.

Thrift, N. (1997) 'The still point: resistance, expressive embodiment and dance', in S. Pile and M. Keith (eds), *Geographies of Resistance.* London: Routledge. pp. 124–51.

Questions about the ways that cities and their spaces are imagined and ordered are important for a critical understanding of the urban and its emancipatory potential. In an age that seems somehow 'after utopia', it might seem odd to be thinking about utopic trajectories. But as David Harvey (2000) argues in *Spaces of Hope*, the rejection of utopianism in recent times has the unfortunate effect of curbing the free play of imagination in the search for alternatives. The utopic trajectories discussed in this section are imaginative spaces or flows through which we can reflect on and critique the urban order and emancipatory politics. The authors look at quite different utopic trajectories – seemingly fixed and stable spaces, the flow of water, the imaginative geographies of films, and the ghost world. They do not seek to create utopian alternatives but to confront the relationship between spatiality and authoritarianism.

Jennifer Robinson asks – Is the emancipatory potential of the city to be found only in those spaces that are characterized by fast movements, seemingly endless mobility, and dynamic mixing? In so doing she makes a case for appreciating the diverse spatialities of urban transformation and emancipation. To the usual list of dynamic inner city sites she adds the seemingly less promising and more stable spaces of fixity, borders, divisions, and territories. She investigates such spatialities drawing on aspects of South African urban political history. Following Lefebvre she argues that seemingly fixed spaces, such as territories, can in fact be transformative because they have an ambiguous continuity with other spaces. The interconnections and flows that make up the spatial reach mean that no space can stand alone. Her practical, political case is that if scholars and policy-makers are to assess the emancipatory potential of South African cities all their different spatialities need to be considered together.

Matthew Gandy focuses on the dialectic between water and urbanism. He argues that water provides a lens through which we can observe the complexities of urban politics, for example, the evolution of different forms of city governance and the incorporation of the human body into the physical fabric of the modern city through the circulatory dynamics of water infrastructure. Outlining the hydrological order and associated discourses that were imposed on the city over time by sanitary reformers and others, he shows how such historic associations between water supply and emancipatory politics were later disrupted by changes in the public realm. The flow of water throughout the city was both emancipatory, in terms of enabling access and sanitation, and yet also politically ambivalent, for it did not seek to undermine the capitalist city. Nevertheless, water remains a utopian trajectory for urban politics, not just for those threatened by thirst and disease in developing world cities, but also as part of the contemporary 'brown agenda' of environmental justice in the developed world.

Geraldine Pratt and Rose Marie San Juan explore the role of urban space as represented by cinema in both the double sense of real space (location) and reel space (connotation). In particular, they are interested in the imagining of utopic urban space in two films, *The Truman Show* and *The Matrix*. In neither film is emancipation fixed in space; rather the heroes both learn to leave the familiar confines of the city as they seek something beyond the confines of the presently imaginable. Critiquing utopic urban form – new urbanism and the modernist city respectively – both movies offer a horizon of possibility that is a non-place from which to reflect on alternatives. This is in direct contrast to Harvey's (2000) discussion of the films *La Haine* and *Deux ou trois choses que je sais d'elle* where he finds no alternative to the place that is capitalist social order.

Steve Pile also takes us beyond the here and now through a discussion of the place of ghosts in the city. He asks whether ghosts can help the living out of the nightmares of the hopeless city. He wonders whether ghosts are figures through which we might understand the unresolved agendas of the past, for the dead generations might have something important to say about our desires for freedom in the city. He is interested in the spectral geographies of the city, in the fissured emotional geographies of pain, loss, injustice, and failure that haunt the city. The ghosts of dead ideals must not, he argues, stop the creation of the ghost of a future yet to come, for now is the time to begin to imagine the emancipatory city.

10 The Urban Basis of Emancipation: spatial theory and the city in South African politics

Jennifer Robinson

This chapter will explore the ways in which the spatiality of cities shapes their role in emancipation or, more narrowly, in political transformation. In the traditions of geographical thought, space has been theorized as landscape, territory or place, as produced space, social space or, most recently as networks of connections and flows. These diverse accounts of space have lead some commentators to suppose that space is a 'chaotic concept' (e.g. Sayer, 1985) or to disavow earlier traditions of thinking about the spatiality of cities in favour of new ideas of networks and cities-as-assemblages (e.g. Graham and Marvin, 2001). In considering the potential role of cities in emancipation, my concern is that by ascribing transformation only to those elements of social life whose spatiality is dislocated (Laclau, 1994), or perhaps interconnected in a network (say, of global flows), we may overlook the transformative potential of other persistent aspects of city life, such as borders, divisions, territories, or the fixity of the built environment, which might, at first glance, seem rather less promising.

This chapter explores the diverse spatialities of urban political change through the experience of South African cities. In South Africa, the city has been a key site in the political struggle against apartheid; it is currently the site of hopes for emancipation from poverty; and for a long time it was the site most feared by the architects of unfreedom and racial injustice: apartheid's perpetrators were most afraid of the African presence in cities. In this chapter I will review some of the ways in which relations between cities and freedoms have circulated in South African history and urban politics. But in the process I also want to say something more general about how it is that city spaces can help to make freedom, to emancipate people from domination and poverty.

Of course, the other side of city life is just as real – cities are also sites and sources of oppression, exclusion, and impoverishment. In the South African context these dimensions of city life have received by far the most attention: how cities were designed to suppress the black majority and enrich the white rulers (e.g. Davies, 1981; Robinson, 1996; Western, 1996), and how city form after apartheid continues to militate against the poor and marginalised becoming part of the mainstream of economic activity and political life (e.g. Mabin, 1995). More recently, commentators have also noted how post-apartheid integrated city planning and developmental ambitions continue to reproduce apartheid inequalities, and even, in the context of a rising neoliberalism, extend them (Bond, 2000).

Apart from the hopes of the anti-apartheid revolution and the forging of a new political order in the wake of the first democratic elections in 1994, the South African city has been associated with anything but emancipation. This context offers something of a caveat then to resting future hopes for emancipation too closely on the city. But even in contemporary South Africa, I argue, the excavation of emancipatory moments of urban life is possible: the potential for urban space to feed into and support revolutionary and transformative moments has been and, I argue, continues to be evident.

The discussion below draws widely on aspects of South African urban history to advance an argument concerning the diverse ways in which urban spatialities shape emancipatory politics. To mark out the ambivalent role of the urban in political change, I will move between three key moments in the history of South African cities: first, a moment in the 1930s, before apartheid, when the potential for realizing the modernist dreams of a capitalist economy in Africa, and for progressive improvement of all people in that country, found some purchase among an African urban elite, while the Eurocentric white population was forging its sense of identity in relation to Empire and racial exclusion. Establishing the grounds for segregation marked a crucial concern among the white elite with the city's potential to popularize emancipation among the African people. Second, I will consider key moments of anti-apartheid struggle in the 1950s and 1980s, when organization around urban issues in modernist townships threatened the roots of apartheid, and crucially depended upon the segregated form of the city. And finally, the 1990s and early 2000s when hopes for emancipating the mass of the African population under a post-apartheid government collided with the neoliberal form of economic engagement with the global economy.

Torn between equality and integration, and heightening economic inequality alongside a de-racialization of the elite, the South African city exemplifies both the potential and the limitations of the city's role in emancipation. The chapter addresses, in turn, aspects of the South African city as a space of assemblage, as deeply territorialized, and as always open to imaginative re-visioning despite the fixity of the physical fabric of the city. The chapter concludes by linking these historical observations of city politics to theoretical understandings of the transformation of urban space, in an attempt to rescue a sense of continuing emancipatory potential from the sometimes gloomy record of post-apartheid policy initiatives.

A troubling cosmopolitanism: cities as spaces of assemblage

As he draws his account of social space to a conclusion, Henri Lefebvre, theorist of space and revolution, makes the following claim:

> The form of social space is encounter, assembly, simultaneity. But what assembles, or what is assembled? The answer is: everything that there is in space, everything that is produced either by nature or by society, either through their co-operation or through their conflicts. Everything: living beings, things, objects, works, signs and symbols. (1991: 101)

If assemblage is a characteristic of spaces in general, it is distinctively the quality of cities: they assemble things, people, and ideas from far beyond their borders. The possibility, then, is that in bringing different things together, cities might produce something new. There have been many accounts of cities as sites of innovation and sources of creative energy. But an important element of the way in which cities assemble things is that they also impose a certain ordering on the people and things gathered together. Being in the same city is no guarantee of encounter, for example, as different land uses, or different kinds of residents, are often deliberately kept apart – and nowhere is this more true than in South Africa. It was the desire to prevent the possibility of new social forms emerging, as well as the chances of unwanted encounters, which lay behind the apartheid city designers' efforts to order urban life there.

In the 1930s, two decades before the formal apartheid city was invented and implemented across South Africa, a vibrant cosmopolitan urban culture was thriving in parts of major cities. African migrants to the city, living in poor and run-down neighbourhoods, and in some of the precursors of apartheid townships, drew on a combination of indigenous traditions, foreign styles, and other South African traditions, to invent new musical styles, produce novels and plays to dramatize their circumstances, and to carve out a distinctive nationalist politics, which was to persist through the century (Coplan, 1985; Couzens, 1985). An Empire Exhibition, staged in 1936 by the business leaders and city fathers of Johannesburg to celebrate a modern, capitalist South Africa emerging from the depression, symbolized something of the way in which the city in South Africa was working at this time (see Robinson, 2003). As one commentator at the time noted in relation to the forthcoming Exhibition:

> A picture offered itself of the tens of thousands of folk of all ages and diverse races who would soon throng this miniature city, peopling the roads and terraces and pavilions, crowding into hours and days a fuller realisation of British Africa and its Commonwealth relationships than could be obtained in a lifetime of office and workshop and field and home. (Crocker, 1936, p. 45)

African people joined the throngs of white South Africans for whom the exhibition was intended to foster a sense of pride in a progressive young (white) nation, anxiously grappling with incipient Afrikaner-English divisions. Africans participated in the exhibition in a range of different ways – indicative of how cities at the time were both assembling and dividing different groups of people. African art was displayed both as part of the general exhibits, and as distinctive ethnographic displays. African performers were part of elaborate stagings of a pageant of South African history. They joined in productions of ethnic dancing (both of these managed by a Belgian stage director, Andre van Geyseghem) but African jazz bands also performed at the evening concerts alongside white and foreign musicians. Perhaps most disturbingly, African people were on display at the exhibition. But many African schoolchildren and adults also visited the exhibition, albeit on specified days and through specified entrances. In these days of proto-apartheid segregation, mingling and interaction were shaped and ordered through formal and informal rules, in this case, ones invented at the last minute by the organizers in conjunction with local native administrators and the prime minister himself.

White enthusiasm for African culture was also evident at the Empire Exhibition and in the city more generally at this time. However, their engagements with African culture were shaded with a form of primitivism, common in European cultures at the time, but in South Africa this was often inflected with a reassuring prescription for tribal segregation. In contrast to this, a proto-middle class African culture, which valued education, literature, and Western style was a striking feature of South African cities through the 1930s (Couzens, 1985).

The cosmopolitanism of South Africa's cities and of the spaces of interaction fostered by them (both by events like the Empire Exhibition and by everyday encounters in streets, homes, and workplaces) was deeply shaped by hierarchies and exclusions. For example, observers noted that some African children at the Exhibition were afraid to enter some of the buildings where displays were held because they assumed they were excluded. Of course some of the poorer white people at the exhibition also found it hard to afford some of the more popular paying attractions. Taken as a whole, the Exhibition highlights the customary practices of the city in which it was held, Johannesburg, where, like many cities, a formal cosmopolitanism was shaped and limited by racial and class-based hierarchies of exclusion. In South Africa these hierarchies hardened over the years into the stark geometries of the apartheid city.

The flows of cultural and political inspiration to South Africa from US African-American intellectuals, from the rights-based legal claims of the British metropole and the white regime, and from indigenous political practices, coalesced in the cities to produce a distinctive and articulate urban African culture and anti-segregationist politics (Coplan, 1985; Nixon, 1994). It was this distinctive and threatening cultural and political milieu, along with the more general response of authorities of the day to the apparent disorderliness of slums, which caused such concern to white residents and the government. In its most extreme forms, white anxieties were translated into urban segregationist strategies, breaking up diverse and creative inner city neighbourhoods, like Sophiatown in Johannesburg, where interactions and freedom from white control had enabled cultural and political innovation to thrive. Over the years, African populations were dispersed to orderly, closely supervised townships, where education, cultural activity and political organization were to be provided or monitored by the white authorities. The freedoms of the city – from places to stage plays, or the freedom to move around to perform music (Ballantine, 1993), or to assemble in protest, were to be closely regulated and controlled.

Anxieties about the disorderliness of the city of assemblage have of course been a common theme among urban reformers all over the world (Boyer, 1983). And while contemporary understandings of the creativity of urban life offer more tolerance of disorder, following interventions by writers like Sennett (1971), the conflictual coming together of differences in the city continues to vex policy-makers, even as they offer opportunities for potentially productive and new kinds of social forms and economic activity to emerge. If, following Massey (1994: 154), 'what gives a place its specificity is not some long internalized history but the fact that it is constructed out of a particular constellation of social relations, meeting and weaving together at a particular locus', then urban policy making is about managing the diverse flows and networks that come together in the space of the city.

In post-apartheid South Africa, making the best of what a city has and encouraging economic growth have become the dominant regulating ambitions of urban governments. But governing in the post-apartheid era is also confounded by the diversity and unpredictability of the social and economic life of cities as spaces of assemblage. Post-apartheid, South African cities have seen numerous new kinds of flows routed through their spaces: migrants from across the continent (both economic migrants and refugees), flows of drugs and chemicals, new directions for guns, private armies, cars and commodities headed for neighbouring countries, and relatively small (but just as influential for their absence) flows of foreign investment capital. South Africa has become an enthusiastic member of global regulating agencies, like the WTO, GATT, and the World Bank/IMF agencies. The ways in which the flows of goods capital and money are regulated, and anticipated, bear all the marks of a self-imposed structural adjustment policy. Not only is the form of assemblage and potential transformation of space and economy in South African cities shaped by the internal ordering and policies adopted by local and national governments, but as global governance is deterritorialised, the ordering of urban space is shaped by much wider flows and policy inputs.

Caught between the vibrant disorderliness of assemblage in the cities and the neoliberal agendas of international and national policies, the South

African city is once again being spatially re-ordered. The differentiation of the spaces of the city – this time, a post-apartheid form of spatial differentiation – is once again shaping the potential that the assembling of diversity might offer. Parts of the city have become home to large numbers of foreign African migrants, many without legal residence status in South Africa, who, consequently, avoid the formal institutions of law and governance (Beall et al., 2002). Although dynamic and internationally connected networks of trade and investment stretch out from these dense concentrations of foreign migrants, there is little ambition among urban governments to provide for or include their activities. Nevertheless there is still at least the potential to sustain their livelihoods, foster their creative activities, and in this way see the cultural and linguistic diversity they introduce as a resource, not a drain.

At the other end of the spectrum of the differentiated post-apartheid city are the locations of international business. No longer in the dense CBD areas that are now the province of back offices, cheap retail outlets and informal trade, they have moved to the weathiest former white suburbs. Very congested as a result of uncontrolled private-sector driven growth, these parts of the city have become the focus of economic policies eager to keep the dynamic formal sector heart of the city's wealth intact. Here the politics of transformation is at its sharpest edge: how can the ordering of the urban economy in these places of privilege build on the creative energy of flows and networks which tie them into economic processes beyond the city, to enhance the well-being of the whole city? Substantial city-wide redistribution is one option. Ensuring that the possibility for mobility and interaction across the different spaces of the city is retained is another. Neither of these is very popular among the wealthy. But if the post-apartheid city is not to repeat the traumas of cultural, political, and economic impoverishment which followed on the territorial divisiveness of the apartheid city, then attention to fostering links, connections, and flows across the city is crucial. The city could become a space of assembly that actually brings differences together in a productive way, rather than keeping them apart.

Nonetheless, I want to argue that there remain important roles for division and territories in an emancipatory urban politics. In a context as historically divided as South Africa, this is a difficult, some might say untenable, position to adopt. But the following section explores how, even here, territorialization and demarcation can play a part in the emancipatory city.

Territorializing social life: is there transformative potential in divisions?

> Visible boundaries, such as walls or enclosures in general, give rise for their part to an appearance of separation between spaces where in fact what exists is an ambiguous continuity. The space of a room, bedroom, house or garden

may be cut off in a sense from social space by barriers and walls, by all the signs of private property, yet still remain fundamentally part of that space. (Lefebvre, 1991: 87)

In South Africa, the city was indeed divided. African townships were clearly demarcated, often distant from the city centre, inaccessible, and even surrounded by walls, fences, or other impassable obstacles (like rivers, railways, or major roads). But there were other dynamics too, which support Lefebvre's suggestion that these divisions are permeable and ambiguous. Flows and movements regularly connected divided parts of the city: workers travelling into the city centre, day after day; goods and services flowing back; images, adverts, radio, and later television; armed police and military personnel. These connections were crucial to both white city and black township. But the borders were there, and the South African city was profoundly territorialized. While these territories were marked and produced out of a complex mixture of ambitions, including racist separation, modernist planning, urban housing delivery, and political control, their effects were not entirely predictable by their architects.

For in bringing together and segregating African and other black populations, the apartheid regime sowed the seeds of its own demise. The territory of the township provided one important context for effective political opposition to apartheid. For example, the nationally significant 1952 protests in Port Elizabeth (South Africa's fourth largest metropolitan region and at the time the second most important harbour in the country) drew on links between communist party members, trade union organizers, and the local ANC branch but also depended upon a 'well established local tradition of mass protest' based in the adjoining township of New Brighton (Lodge, 1983: 55). Important here were longstanding histories of township-wide protests over rent, housing conditions, and council regulations. The 'native administration' responsible for managing the township even encouraged mass meetings of the local population to discuss these grievances and regularly received petitioners and protestors in the offices (Robinson, 1996). Similarly, during the final round of political opposition to apartheid in the 1980s, it was the alliance of township-based urban movements that eventually brought down the apartheid government. In this way the demise of apartheid was inadvertently precipitated by the territorial divisions it had implemented to shore up its power.

But how would a progressive form of contemporary city politics make productive use of territorialization, differentiation, and division in the city? For instance, ward and community-based forms of representation are common to the functioning of urban democracy, while keeping dangerous and unhealthy activities away from where people live and children play is a demand shared by rich and poor neighbourhoods alike and a key ambition of local governance. Some kinds of differentiation sustain inequality, of course.

In practice, zoning often serves to resolve urban contradictions in favour of middle class home owning interests (Boyer, 1983), but this result is by no means inevitable and forms of differentiation can also be productive and benign in an urban context.

One of the key features of territorialization is its fundamental contestability. Like all aspects of urban space, borders and divisions are subject to transgression by the objects and people they aim to exclude – what Laclau (1994) would call their 'constitutive outside'. Moreover, as urban space changes, so the meaning and impact of different divisions can also change profoundly.

This contestability is especially crucial to understanding the potential roles of division in formal political representation. During the protracted negotiations around remapping the institutional and electoral form of the apartheid city, the contested and unpredictable consequences of territorial divisions proved enabling for political transformation. The demarcation of city borders and the internal divisions of the city into wards for electoral representation have important political consequences for municipal government, where choices about where to draw a boundary potentially can determine which political party will represent an area. Considerations of partisan politics were compounded in South Africa by temporary transitional compromises to weight white votes more heavily than black ones (see Robinson, 1998, for details). In general, it was correctly assumed that whites would vote for the former ruling National Party (NP), blacks for the African National Congress (ANC). Other racial groupings showed somewhat unpredictable affiliations. Coloureds in the Western Cape and Indians in Durban were expected to vote for the ANC, but for different reasons, many of these voters opted for the NP and other minority parties in the 1994 and 1999 elections.

Given the racialized form of the apartheid city and of political affiliations, the local geography of electoral districting became the object of strong political contestation. Finding a balance between representing local community interests through ward demarcation and ensuring city-wide representation of the overall balance of political and racial affiliations, proved a serious challenge at this stage of South Africa's history. The process of re-territorializing the post-apartheid city brought to the fore the fundamental contestability of urban space. Importantly for South Africa's transition, the conflict over redistricting was resolved through democratic and legal procedures, which contributed to ensuring the implementation of new, representative forms of local government.

In post-apartheid cities, new forms of territorialization are also playing an important role in plans for economic growth. Theorists of the urban economy argue that many of the flows and networks that make up city economies work best when clustered and benefiting from the synergies of co-location in neighbourhoods or districts with similar or related activities (Storper, 1997).

Promoting these agglomeration economies requires planning for differentiation and specialization. All too often, however, spatial plans in urban South Africa have resulted in the removal of some of the poorest people in the city to accommodate grand development schemes designed to produce better opportunities for the city as a whole. Recently, hundreds of squatter families were evicted from an abandoned 'Turbine Hall', previously used for electricity generation, to make way for a new cultural district in that rather run-down part of downtown Johannesburg. The plan was to use the land for a shopping centre and other attractions for the cultural district, alongside existing museums, music venues and art galleries (Dirsuweit, 1999). A lot could be gained for inner city Johannesburg by capitalizing on these assets, facilitating the emergence of creative intersections within this part of the city. But the tactics required to achieve it involve demarcating different zones of the inner city for different activities and enforcing those divisions. Territorialization may be productive, then, economically, but it also calls attention to the contested politics and power imbalances involved in trying to negotiate improvements in city life.

One final form of territorialization I will mention here is potentially productive of economic growth in poor countries in general, and in South African cities too. The territory of the city as a whole has taken on a greater significance in economic planning, both in South Africa and in other poor countries where democratization and decentralization have increased local government autonomy. The World Bank (2000) and its various urban partners are promoting the idea of formulating city-wide development strategies through a process which would bring together the diverse interests present in any particular city to create a vehicle for planning the economic future of cities. These new policy initiatives have an ambivalent edge. On the one hand, they could be read as the implementation of structural adjustment policies for newly democratised urban governments, struggling to build capacity in this new area of economic development and now available as clients for the international agencies seeking to justify their existence. On the other hand, though, they also offer an opportunity to consider the diversity of city economies, to respond to the specific economic mix of particular urban areas, and to negotiate plans for the future of cities across the conflicting interests and differentiated zones of the urban arena. More hopefully, a city-wide perspective is essential if redistribution is to play a role in poverty alleviation (UNCHS, 2001).

The different aspects of the territorialization of cities and city economies which I have outlined in this section are an important component in understanding the nature of political transformation. The potential for economic transformation and redistribution in South Africa's cities in the future will certainly be shaped by these territorializations and, in some instances, enhanced by this form of spatiality. In that sense territories and divisions are a crucial part of any account of the emancipatory city.

Change or stasis: the dynamism of dreams and imagination

A revolution that does not produce a new space has not realized its full potential; indeed it has failed in that it has not changed life itself, but has merely changed ideological superstructures, institutions or political apparatuses. A social transformation to be truly revolutionary in character, must manifest a creative capacity in its effects on daily life, on language and on space. (Lefebvre, 1991: 54)

One of the most frequently used tests for the effectiveness of political transformation in South African cities is the extent of change in urban spatial form. And if Lefebvre is to be believed, in the quotation above, then South Africa has not progressed very far in its hopes for effecting revolutionary change in its cities.

Observers have noted that the apartheid-era urban form is likely to be very persistent (Simon, 1989; Western, 1996). Townships remain distant from major nodes of economic opportunity, requiring long commuting distances on still unsafe and costly modes of transport. Far from addressing this problem, post-apartheid housing developments, combined with the continuation of informal squatting, reinforce the location of the poorest in areas on the periphery of the city (Mabin, 1995). Bond (2000) notes that reliance on private sector delivery and financing for basic services and housing for all but the very poorest has meant that new housing, for example, is as bad as, and often considerably worse in terms of structure and finishing, than the matchbox houses of the apartheid era. Inadequate controls over delivery mean that many communities are being deprived of housing access as subsidies are directed by developers and municipal authorities towards costly infrastructure and services rather than the houses themselves. The costs of acquiring land in the central areas of cities, and the alternative development potential of these parts of cities, mean that poor residential areas are still being developed on the urban fringes. Informal areas are usually even further out, although there are many examples of small squatter settlements in vacant land and along transport routes closer to city centres.

In the face of large-scale pro-privatization initiatives and the strong influence of World Bank thinking on neoliberal urban governance, South African cities are set to entrench fundamental class divisions, which historically follow racial lines. Combined with the retreat of some middle and upper class people to defended security villages (Hook and Vrdoljak, 2002), the transformation from apartheid to post-apartheid starts to look rather like a deracialized recasting of entrenched divisions (Saff, 1994). Indeed, many observers consider that perilously little has changed in the post-apartheid era, and the persistence of urban segregation, albeit increasingly on a class basis, is seen as some of the strongest physical evidence of stasis, as opposed to real political and economic change.

Personally, I find this analysis both convincing, and not so. The achievements of political transformation are substantial. While a predominantly African ANC government in Johannesburg may choose to harass survivalist street traders in favour of cleaning up the city, the democratic achievement of the transition in creating such a government should not be underestimated. Similarly, housing delivery to the more than 1 million households given subsidies since 1994, must count for something. More hopefully not all international policy inputs are purely neoliberal. A strong pro-poor service delivery agenda circulates in the city of Johannesburg as a result of inputs from a range of research and donor agencies. But substantial levels of historical debt and low recovery of service charges have left this city with a major need for institutional reform, even without the demands of post-apartheid redistribution. Thus, the neoliberal restructuring of local government in Johannesburg, labelled iGoli 2002 (after the city's vernacular name, meaning place of gold), has been crucial for, 'without success in the technical transformation of local government, the pro-poor goals of reconstruction will not be achieved, and this is the fundamental premise of iGoli 2002' (Beall et al., 2000: 118). This said, the challenges of transformation are indeed multiple – institutional, fiscal and redistributive – and cities like Johannesburg have many constituencies and ambitions which need to be balanced, or whose priorities have to be negotiated and contested in what are necessarily conflictual city-wide political forums.

Despite the best efforts of the authorities and communities, some elements of the segregated apartheid city are unlikely to disappear, and may well be entrenched through new political and economic dynamics. We may want to ask, however, whether re-arranging the spaces in question is really necessary for political transformation. Perhaps the important thing would be to imagine different types of connections and flows shaping the opportunities and quality of life for people living in segregated areas? Instead of changing the location of physical things and the visible arrangements of space, we might transform the spatial relations and networks into which they are connected, creating, as Dewar (1995) suggests, 'extroverted' spaces, that look outwards to the opportunities of the wider city, rather than inwards to the self-contained zones of the apartheid city.

In this regard, it is worth considering one final and all too easily forgotten question: in the face of an unchanging physical environment, what is the potential for emancipation through the imaginative reconfiguration of social space? Here, too, we can find some direction for our analysis in Lefebvre and his critical consideration of a 'psychoanalysis of space' (see Pile, 1996, who establishes this reading of *The Production of Space*).

Alongside spatial practice (how space is produced and used in particular societies) and representations of space (conceived space, the space of scientists, planners, urbanists, etc.), Lefebvre (1991) described representational space. He defined this as 'space as directly lived through its associated images and symbols, and hence the space of "inhabitants" and "users", but also of some artists and perhaps of those, such as a few writers and philosophers, who *describe* and aspire to do no more than describe. This is the dominated – and hence passively experienced – space which the imagination seeks to change and appropriate.' (Lefebvre, 1991: 39). Here in representational space, then, is a form of understanding spatiality that draws on cultural and historical resources, the possibility and memory of other ways of living in spaces. Lefebvre's ambivalent attitude to this space as a source of change is indicated by his dismissal of writers, artists, and so on as simply describing, rather than actually bringing about a transformative intervention into the production of space. This ambivalence regarding the emancipatory potential of language and imagination runs through his rather testy engagement with psychoanalysis and semiotics (Pile, 1996: 148).

But as Lefebvre seeks to determine the qualities of this source of potentially new spatialities, the body and the unconscious come into view to signify the possibility of spaces beyond the dominant, abstract space. The body is drawn on because it is a lived, sensual realm beyond and potentially disruptive of the visually based orderings that underpin abstract, or conceived, space. The unconscious also signifies this possible other space because it is a space of 'affectivity' that cannot be reduced to the rationality of abstract space. In the same way as the unconscious represents (an element of) the subject that is at the same time present and denied, Lefebvre suggests that the realm of affectivity stands in a similar relation to abstract space. The unconscious supposedly consists of material that cannot be symbolized and is formally excluded from the conscious world of language. The rationality and symbolism of abstract space similarly cannot make sense of the realm of affect and the lived body; it excludes them from and subordinates them to the ordering regimes of the visual, the geometric, the planned. It is to these excluded, subordinated, repressed realms of emotions, affect, and physical sensuality that Lefebvre looks for one potential means of transforming space:

> social space ... contains potentialities – of works and of reappropriation – existing to begin with in the artistic sphere but responding above all to the demands of a body 'transported' outside itself in space, a body which by putting up resistance inaugurates the project of a different space (either the space of a counter-culture, or a counter-space in the sense of an initially utopian alternative to actually existing 'real' space). (1991: 349)

The subject moves in space, experiencing and remaking the meaning of spaces initially constituted to speak of power. In their everyday activities, subjects are witness to the possibility of other forms of spatiality through their bodies and their movements, as well as in their imaginations, in the dynamism of their inner worlds which are both made through, and remaking of, the 'external' spaces of the environment (Pile, 1996). The body, the unconscious and our inner worlds clearly play an important role in the production of the meaning of space, and in its potential transformation.

Lefebvre's notion of representational spaces provokes a sense of how we might imagine spaces being transformed, even in the absence of substantial physical changes. Every time we move around in the city we potentially use spaces differently, imagine them differently. Different people in the city have different resources to draw upon in their imaginative re-use and re-making of the city – different histories, different positions. From this perspective the creative potential embodied in the diverse populations of South Africa's cities appears vast. Even in conditions, then, when the urban environment seems to be fixed and its materiality relatively unchanging, there may be the possibility for imagining and shaping new kinds of spaces, for finding spaces transformed, moved, shifted into strange-ness, as if in a dream …

So while planners and government try to find ways to re-arrange the city, or to encourage re-connection and integration, ordinary people are re-using and re-making urban space at a rapid rate! Faced with a restricted local market in poor townships, many informal sellers have relocated to central city areas. The streets of downtown Johannesburg, for example, now primarily a retail and service centre for outlying Soweto, are covered with informal stalls set out in the shadows of the tall skyscrapers, selling everything from fruit to handbags, cosmetics and clothes, as well as haircuts, shoe repairs, photographs and telephone calls. The major retail stores continue to trade behind the facade of street-sellers. The symbolic spaces of the capitalist class, the verticality of dominant spatialities, sit juxtaposed with all kinds of people walking along and using the pavements of the modern city, making it, experiencing it and imagining it different(ly). Some traders find it necessary to stay on the streets with their goods overnight and in some places street settlements have sprung up, allowing easier access to the centre than do distant townships. There is also a number of more central areas where squatters have seized the opportunity to build more permanent shelters in a well located area, rather than on the more remote, if more abundant land on the edges of the city where most newcomers and homeless people are forced to seek shelter. The lines of the city are crossed, redrawn, reimagined – outside of the conceived spaces of planning visions. The city of everyday experience and imagination is already a different space: it is already a space of difference, or perhaps even 'differential space', which Lefebvre hopes will eventually succeed the abstract space of capitalism.

Planners do well to keep up with peoples' energetic redrafting of city space. The more progressive city councils are following emergent concentrations of traders with stalls and facilities, arranging transport services in response to new concentrations, offering services and subsidies in the wake of a land occupation. This means learning from the creative re-use of space rather than trying to reassert formal and rational ideas about the proper use of certain kinds of spaces. Utopian post-apartheid city plans are anyway made a nonsense by government decisions (developer-led subsidy schemes locate housing on the fringes of the city – the dream of compaction and integration undermined by pragmatics and neoliberalism) as much as by ordinary people's activities. A new de facto relationship seems to have emerged between planning visions driven more by a dominant abstract spatiality, and the spatial practices and representational spaces of everyday life (Dierwechter, 2001). Is it possible to claim that out of these interactions, urban space has been transformed?

As Lefebvre (1991: 42) searches for ways of imagining the transformative (representational) spaces of everyday imaginative re-appropriation of space, he places in this vital component of the dialectical and dynamic production of social space, phenomena such as 'childhood memories, dreams, uterine images and symbols'. The dream is perhaps one of the more useful metaphors to use in order to understand this aspect of the politics of urban space, for as a representation of the dynamic agency of the unconscious, its symbolization of unconscious feelings is only ever partial. Locked into the form of the picture, muted by the 'repressive agency' and working in the terrain of the unspeakable, the dream is never an uncompromised 'other' space. In Freud's account, the dream is always mediating the energy of the unconscious with the caution of the conscious self, the ego, in the interests of not startling us from our sleep. In the same way, we could imagine new spaces in the city guarded and muted by the spaces of the old order, for whom our dreams may be their worst fears. Even so, dreams of new spaces may not be able to leave behind elements of the past; the spaces people dream of creating may always drag along with them the restricting voice of past times. Planning dreams, for example, speak in the only language they know: a rationalizing, physicalist perspective on the internal relations of the city (Dierwechter, 2001). But there remains the possibility of re-viewing those spaces, seeing them differently, and trying not to bring them into the sharp line of vision that has always enabled dominant orderings of space. Imaginations of urban futures and of the transformative potential of cities could stretch to draw in the everyday, affective, and sensual experiences constantly remaking the meaning of the city.

The practical consequences of this aspect of urban spatiality may be slight: taps alongside street traders; permissive use of sidewalks; marked taxi ranks; diverse cultural interpretations of city monuments. And indeed, even those achievements can be undermined as councils may (and do) implement services in neoliberal and excluding ways: for example, making street traders pay for market places. But as an analytic of the emancipatory potential of cities, the dynamism of everyday imaginative and practical re-uses of urban space is a crucial component. Moreover, I would argue that this is a crucial component of understanding the potential for transformation in poor cities dominated by informality and lacking institutional resources to effect sustained and highly visible physical changes in urban spaces (Simone, 2001). It also has pertinence in wealthy cities where the politics of exclusion means that city spaces produced in the image of dominant classes and social groups must find their meaning for many residents beyond the semiotics of the dominant culture.

Conclusions: the many spaces of urban emancipation

The transformation of post-apartheid South African cities has entailed a range of different spatialities. From mobilizing the territorialized communities of the segregated city to redemarcating the political divisions of cities; from literally transforming the landscape with the construction of many more houses and residential areas to the racial and cultural transition of the character of different suburbs; the South African city has certainly been changing. These cities are also linked into networks with a range of different spatial reaches – from African migrant trading links around the continent, to transnational firms based in South Africa but spreading their investments and influence far beyond city borders. The spatiality of connections and flows is clearly also a source of transformation.

In this chapter I have made a case for appreciating the diverse spatialities of urban transformation and for including accounts of the politics of boundaries, territories, and spatial stasis, alongside those of interconnections and dislocation, in any analysis of the urban basis of political transformation.

Whether these varied spatial sources of urban change in South Africa amount to emancipation is a trickier question to address, and commentators are divided. There is no doubt that the territorialized and urban basis of political organization played a huge role in the overcoming of apartheid, especially through the 1980s, and clearly enabled the political emancipation of South Africa. It is less clear in which direction the many different urban sources of political change are now heading. Many observers are deeply pessimistic about the possibility of realizing anti-apartheid hopes of emancipation in the current neoliberal policy climate. But what I have argued here is that if scholars and policy-makers are to assess the emancipatory potential of these cities, even in these less hopeful times, the full range of spatialities needs to be considered together. Each of the spatialities of the city I have outlined here has a role to play in emancipation and political transformation. Territories, divisions, and the dynamic worlds of imaginative and embodied subjects, just as much as networks, connections, and the dislocations of political spaces all contributed to transforming South African cities in the past, and will continue to do so into the future, in South Africa as elsewhere.

References

Ballantine, C. (1993) *Marabi Nights. Early South African Jazz and Vaudeville.* Johannesburg: Ravan Press.

Beall, J., Crankshaw, O. and Parnell, S. (2000) 'Local government, poverty reduction and inequality in Johannesburg', *Environment and Urbanisation,* 11: 107–22.

Beall, J., Crankshaw, O. and Parnell, S. (2002) *Uniting a Divided City: Governance and Social Exclusion in Johannesburg.* London: Earthscan.

Bond, P. (2000) *Cities of Gold, Townships of Coal*. Trenton, NJ: African World Press.

Boyer, C. (1983) *Dreaming the Rational City*. Cambridge, MA: MIT Press.

Coplan, D. (1985) *In Township Tonight! South Africa's Black City Music and Theatre*. Johannesburg: Ravan Press.

Couzens, T. (1985) *The New African: The Life and Work of H.I.E. Dhlomo*. Johannesburg: Ravan Press.

Crocker, H. J. (1936) 'The Empire Exhibition, South Africa, 1936–1937', *The Almanac*, October: 45–50.

Davies, R. (1981) 'The spatial formation of the South African city', *GeoJournal*, Supplementary Issue, 2: 59–72.

Dewar, D. (1995) 'The urban question in South Africa: the need for a planning paradigm shift', *Third World Planning Review*, 17: 407–20.

Dierwechter, Y. (2001) 'The Spatiality of Informal Sector Agency'. PhD Dissertation, Department of Geography, London School of Economics.

Dirsuweit, T. C. (1999) 'From fortress city to creative city: developing culture and the information-based sectors in the regeneration and reconstruction of the greater Johannesburg area', *Urban Forum*, 10: 183–213.

Graham, S. and Marvin, S. (2001) *Splintering Urbanism: Networked Infrastructures, Technological Mobilities and the Urban Condition*. London: Routledge.

Hook, D. and Vrdoljak, M. (2002) 'Gated communities, heterotopia and a rights of privilege: a "heterotopology" of the South African security park', *Geoforum*, 33: 195–220.

Laclau, E. (1994) *Reflections on the New Revolutions of Our Time*. London: Verso.

Lefebvre, H. (1991) *The Production of Space*, trans. D. Nicholson-Smith. Oxford: Basil Blackwell.

Lodge, T. (1983) *Black Politics in South Africa since 1945*. Johannesburg: Ravan Press.

Mabin, A. (1995) 'On the problems and prospects of overcoming segregation and fragmentation in Southern Africa's cities in the postmodern era', in S. Watson and K. Gibson (eds), *Postmodern Cities and Spaces*. Oxford: Basil Blackwell. pp. 187–98.

Massey, D. (1994) *Space, Place and Gender*. Cambridge: Polity.

Nixon, R. (1994) *Homelands, Harlem and Hollywood: South African Culture and the World Beyond*. London: Routledge.

Pile, S. (1996) *The Body and the City: Psychoanalysis, Space and Subjectivity*. London: Routledge.

Robinson, J. (1996) *The Power of Apartheid: State Power and Space in South Africa's Cities*. Oxford: Butterworth Heinemann.

Robinson, J. (1998) 'Spaces of democracy: re-mapping the apartheid city', *Environment and Planning D: Society and Space*, 16: 433–78.

Robinson, J. (2003) 'Johannesburg's 1936 Empire Exhibition: interaction, segregation and modernity in a South African city', *Journal of Southern African Studies*, 29, 3: 759–89.

Saff, G. (1994) 'The changing face of the South African city: from urban apartheid to the deracialization of space', *International Journal of Urban and Regional Research*: 18: 377–39.

Sayer, A. (1985) 'The difference that space makes', in D. Gregory and J. Urry (eds), *Social Relations and Spatial Structures*. London: Macmillan. pp. 49–67.

Sennett, R. (1971) *The Uses of Disorder: Personal Identity and City Life*. London: Allen Lane.

Simon, D. (1989) 'Crisis and change in South Africa: implications for the apartheid city', *Transactions of the Institute of British Geographers*, 14: 189–206.

Simone, A. (2001) 'Straddling the divides: remaking associational life in the informal African city', *International Journal of Urban and Regional Research*, 25: 102–17.

Storper, M. (1997) *The Regional World*. New York: Guilford.

UNCHS (United Nations Centre for Human Settlements) (2001) *Cities in a Globalising World: Global Report on Human Settlements 2001*. London: Earthscan.

Western, J. (1996) *Outcast Cape Town*. Berkeley, CA: University of California Press.

World Bank (2000) *Cities in Transition: World Bank and Local Government Strategy*. Washington, DC: World Bank.

11 Water, Modernity and Emancipatory Urbanism

Matthew Gandy

The flow of water through urban space has played a pivotal role in freeing cities from disease, squalor and human misery. The construction of intricate networks of pipes through the modern city gradually released urban populations from the daily drudgery of fetching water. And the diffusion of plumbing technologies has brought the pleasures of bathing into the private realm of the modern home. From the intimate space of the bathroom to the hidden space of the sewer, from the civic display of the fountain to distant dams and reservoirs, water provides a link between the corporeal experience of space and the abstract dynamics of capitalist urbanization. Yet this hydrological dynamic is not restricted to the modern city: archaeological and historical evidence abounds of complex water engineering projects in the past. What is distinctive about the modern city, however, is the changing interaction between water supply, capitalist urbanization and the politics of the body reflected in a series of developments ranging from the social production of domestic space to emerging regulatory responses to disease.

Over the last 200 years we can identify three broad phases in the hydrological dynamics of the modern city: the nineteenth-century 'organic city', the twentieth-century 'bacteriological city', and the contemporary 'fragmentary city'. The nineteenth-century association with the 'organic city', for example, involved a circulatory conception of urban space in which the principal concerns were with stagnation, confinement and the loss of valuable organic matter. By the middle decades of the nineteenth century a purificatory discourse began to emerge in the polluted industrial cities of Europe and North America which would evolve into a bacteriological conception of public works to be managed by technical elites. The twentieth-century 'bacteriological city' saw the formal inclusion of advances in medical epidemiology into modern strategies for the protection of public health. A hygienist emphasis on the scientific management of urban space marked a decisive shift in the scope and mechanisms of municipal governance and was at various times combined with a reformist political agenda. By the 1960s and 1970s, however, the emphasis on centralized forms of universal state provision began to encounter a range of fiscal and ideological challenges. In the emerging 'fragmentary city' we find a dislocation between public health and urban infrastructure driven by an increasingly individualized conception of civil society as an amalgam of consumers rather than a polity of citizens. Yet if we apply this urban schema to the colonial and post-colonial cities beyond the core metropolitan regions of the global economy we find that the historic association between potable water provision and the nascent public sphere is far more problematic than it may at first appear. And while the organic ideal may have receded as an integrated conception of urban governance since the nineteenth century we find that key elements of the pre-bacteriological city persist through the networks of wells, pit latrines and other small-scale water technologies that characterize many urban spaces in the developing world. What this historical periodization does suggest, however, is that the relationship between water and cities is characterized by multiple and often contradictory modernities expressed through the social and political dynamics of urban infrastructure.

How can we begin to conceptualize these changes and their implications for urban life? One of the few attempts to apply social theory to the hydrological dynamics of modern cities is provided by the sociologist Thomas Osborne who contends that the development of modern practices of sanitation and public health form an integral element in a liberal political discourse founded on the need for security within the context of capitalist urbanization (Osborne, 1996). The word 'security' in this context is used to encompass the protection of the body from disease as well as wider preoccupations with the maintenance of social order. The process of urban transformation during the nineteenth-century generated immense challenges to prevailing laissez-faire political thought and played a key role in reformulating the scope and rationale of governmental institutions. Yet it would be misleading to conceive of responses to laissez-faire ideology as necessarily residing outside of, or in opposition to, liberal economic orthodoxy: the challenge was to modify aspects of the urbanization process without undermining the prospects for further capital accumulation. An emphasis on the critical role of urban infrastructure in the rationalization of modern societies enables us to find a point of conceptual engagement between sanitary engineering as a technical discourse and the wider dynamics of public health as a fundamental criterion for the maintenance of urban cohesion (see also Armstrong, 1993; Gordon, 1991; Rabinow, 1982). The incorporation of the human body into the physical fabric of the modern city via the circulatory dynamics of water infrastructure illustrates the extent to which new forms of government or 'governmentality' have shaped the structure of urban space. Access to water is a critical dimension to contemporary urban discourse and illuminates wider concerns over citizenship rights, democratic participation in urban affairs and the restructuring of the public realm.

The decline of the excremental metropolis

In the pre-modern era drinking water was typically acquired through wells, springs and water vendors. Limited sewer systems were built to carry storm water away from larger streets and thoroughfares. More sophisticated urban societies such as pre-colonial Delhi or the city-states of renaissance Italy built complex hydraulic engineering projects encompassing aqueducts, canals, underground conduits and other elaborate structures. Imperial Rome, for example, had hundreds of public fountains, baths and water jets in addition to thousands of private bathing complexes. Architectural distinctions between private and public space were quite different from modern conceptions of urban design. In medieval Europe, for example, the use of water for washing remained predominantly a collective activity and often served a recreational or therapeutic role. By the sixteenth century, however, the social and sensual aspects of bathing began to wane as nakedness was recast as a sin. During the Reformation bathing became associated with licentious behaviour and a new bashfulness emerges in relation to the human body. The rediscovery of the pleasures of bathing in the eighteenth century carried with it an erotic charge and often drew on Orientalist conceptions of sensuality derived from European travellers' encounters with the *hammams* of North Africa and the Middle East (see, for example, Giedion, 1948; Vigarello, 1988).

With the growth of industrial cities the relationship between water and urban form began to change rapidly. Emerging patterns of social stratification became enmeshed in the changing economic dynamics of the urban economy. The growing economic disruption and health menace of inadequate urban infrastructure led to new conceptions of sanitary engineering. The problem of cleanliness began to assume greater social and political significance at a time when intensified processes of class stratification found their physical expression in new geographies of squalor and abjection. The use of water was subject to a plethora of competing conceptions of health, hygiene and hydraulic design. The circulatory discoveries of William Harvey and other pioneers of eighteenth-century medical science informed an emerging organicist doctrine that a healthy city required the constant movement of air, water and waste products in order to ensure the satisfactory functioning of the city's digestive and alimentary systems (see Corbin, 1986; Sennett, 1994). The earlier emphasis on the ornamental and sybaritic qualities of water was gradually modified by a strategic empiricism dedicated to the improved management of cities. A hygienist discourse emerged in which the problems of spatial and social purification were seen as different facets of a new moral and political order. A reformist approach to the politics of water gradually emerged in Europe and North America that was predicated on two elements: a rights-based demand for piped water supplies to be extended to the urban poor and an admonitory agenda to improve 'domestic habits'. As water became more a matter of day-to-day survival than in the past the politics of water became increasingly enmeshed in new forms of social and spatial polarization in industrial cities. At the same time, however, that new conceptions of cleanliness and bodily hygiene became established the sanitary conditions of many towns and cities were in a state of rapid decline. This posed an acute dilemma for emerging hygienist discourses that pitched new forms of bodily refinement and spatial organization against the growing poverty and chaos of the nineteenth-century city.

The problem of water-borne disease in the pre-bacteriological era was focused principally on confined spaces or 'diseased tissues' where there was limited circulation of air and water. The miasmic conception of disease epidemiology informed the circulatory dynamics of urban design and provided a powerful focus for institutional architecture. A significant pretext for nineteenth-century urban planning involved the drainage of marshes and the elimination of inner-city slums and rookeries, which were perceived as breeding grounds for disease, crime, and political insurrection. Yet this disciplinary architectural code did not engage with the inherent limitations of existing modes of urban governance. As the spatial problematic of disease became more explicitly an issue of urban governance, the focus shifted from architectural innovation to institutional reform. The problem was not only a matter of strategic coordination between different municipal responsibilities but also the replacement of inadequate private provision. If we survey the changing pattern of water provision in Europe and North America during the nineteenth century we find a decisive shift from private to public supply systems (see, for example, Jacobson and Tarr, 1994; Melosi, 2000). The provision of water supply emerged as a rationale for the development of new forms of municipal governance and marks a radical extension of the power of the state within the context of prevailing laissez-faire conceptions of public policy. In the case of New York, for instance, the completion of the city's new water system in 1842 was marked by the biggest public celebrations since American independence. The technical and logistical complexity underpinning the modernization of urban infrastructure in New York had profound consequences for not only the reshaping of the physical space of the city but also for a nascent public sphere in urban affairs (Gandy, 2002: 32–4).

It is easy to overlook the extent to which the organic ideal differs from later conceptualizations of urban form and thereby overstate the continuities between different phases in the evolution of the modern city. In the case of Second Empire Paris, for example, Baron Haussmann fiercely resisted the connection of newly-built storm water sewers to individual homes. Indeed, the reconstruction of nineteenth-century Paris, so central to many accounts of the origins of modern urbanism, predates the bacteriological transition in terms of the regulatory scope of municipal governance and the technical conceptualization of urban infrastructure (Gandy, 1999). The organic ideal for sanitary reform was predicated on an aversion towards the waste of valuable human manure. The 'frenzied utilitarianism' of the nineteenth-century hygienists called for a physiological conception of urban order that conflicted with the abstract dynamics of capitalist urbanization (Laporte, 2000: 123). Writers as diverse as Jeremy Bentham, Pierre Leroux and Friedrich Engels all expressed concern at the loss of valuable nutrients through the new sewer systems of European cities: Bentham directed that every *'besoin* ... should be put to use as manure'; Leroux railed against Malthus with his contention that the genuine Law of Nature is 'a natural circle or *circulus*' derived half from production and half from consumption (cited in Laporte, 2000: 119, 129); and Engels longed for a utopian city in which there would be a synthesis of urban *Kultur* and rural *Natur* (see Schorske, 1998). Prominent public health advocates such as Edwin Chadwick insisted on a distinction between 'fresh' and 'putrid' sewage and the need to ensure that human manure was not being wasted: not only was this an example of economic inefficiency warned Chadwick but the disposal of human waste through London's sewers also contributed towards the filthy state of the Thames. As late as 1885 Chadwick complained that too much water was being 'pumped into London' and that this was reducing the value of human manure for agriculture (Chadwick, 1885: 37). Yet Chadwick's utilitarian plans conflicted with the rationalization of bodily functions in the modern city. The reconstruction of subterranean London under Joseph Bazalgette and the Metropolitan Board of Works did not follow an organic ideal and opted instead for a combined sewer system which would in turn lead to the development of modern water purification plants and the integration of modern discourses of pollution control into the physical fabric of the city.

The organic ideal owes much to the communitarian rationalism of Fichte, Voltaire and other European thinkers who conceived of the modern city as a centre of freedom, commerce and cultural advancement (Schorske, 1998). They longed for a modern equivalent of Renaissance urbanism where formal geometric ideals in city planning could be combined with a democratic and progressive urban culture. The nineteenth-century hygienists sought to extol the virtues of human excrement as a superior source of manure for agriculture but their enthusiasm for shit conflicted with the relegation of defecation to the private realm of the modern city. Their circulatory and utilitarian conceptions of urban space could not contend with either changing social habits or the continuing threat of disease. The impetus of social and technological change overwhelmed the intricate utopian ideals of an organic urban order and necessitated a series of changes in the scope and mechanisms of municipal governance.

Scientific governance and the interstices of modernity

The 'organic city' proved to be a transitional and ultimately chimerical urban form for the rapidly growing cities of Europe and North America. Increasing water usage in individual homes quickly overwhelmed existing sanitary arrangements and destroyed the economic value of human waste as agricultural fertilizer. Changing attitudes towards cleanliness and human waste, combined with new trends in water consumption and interior design, instituted a shift towards modern plumbing systems. New evidence of the links between disease and drinking water contamination in advance of the bacteriological identification of specific pathogens undermined the technical and political faith in the existing hydrological order. The technological advances that had brought mass-produced water technologies to the modern home occurred in combination with new developments in the petrochemical industries that challenged the dominance of waste-based fertilizers. Advocates of modernist urban design such as Adolf Loos and Sigfried Giedion celebrated rising water consumption as a dynamic symbol of modernity and had no interest in the Arcadian geometries of the organic city. Abundant water use was from the outset a vibrant element in modernist conceptions of urban design. The pristine lidos, state-of-the-art bathrooms and other water features incorporated into modernist architecture celebrated the interaction between water and the human body (see, for example, Braham, 1997; van Leeuwen, 1998).

The emergence of the 'bacteriological city' had profound consequences for the political economy of urban space. In the case of Hamburg, for instance, the devastating cholera epidemic of 1892 exposed the mercantile dominance of city government: 'It demonstrated, with a graphic and shocking immediacy, the inadequacy of classical liberal political and administrative practice in the face of urban growth and social change' (Evans, 1987: 565). In addition to the technological reconstruction of Hamburg in order to provide safe drinking water, the political legacy of cholera also contributed towards the reform of suffrage laws and the reconfiguration of municipal governance. The politics of water played a significant role in the redefinition of urban citizenship and the emancipation of the poor from the capricious self-interest of urban elites. What we are really dealing with here is two phases in the history of liberal political traditions: an earlier phase rooted in a limited redefinition of the public realm and a later social-democratic phase driven by a shift in the political bargaining power of organized labour. The political transformation of nineteenth-century cities such as Hamburg marks a transition towards the modern discourses of pollution control and the creation of the bacteriological city as a distinctive social, political and scientific phase within the history of capitalist urbanization. Yet the emancipatory dynamic of modern urbanism rests on a fragile and only partially extended conception of the public sphere. If we explore the universalist pretensions behind the scientific management of modern cities in a wider geographical context the picture becomes distinctly problematic. In colonial Bombay, for example, the restriction of adequate sanitation to European enclaves within the rapidly growing city contributed towards a series of devastating plague epidemics (Klein, 1986). Far from being a nineteenth-century engineering anomaly, however, this dual discourse of human well-being and sanitary tolerance has reemerged in the contemporary city where the link between the politics of public health and the protection of society as a whole has become increasingly tenuous.

Advances in epidemiological science during the second half of the nineteenth century provided the basis for a new form of social regulation in which moral discourses could be partially displaced by a technical emphasis on the mechanisms of public health improvement. Changing cultures and practices of water use reflect this transition most strikingly through the emergence of the modern bathroom. With the diffusion of the bathroom we find that the corporeal and technological dimensions to private space become subsumed within a wider structure of social control and the regulation of bodily functions (Benton, 1999; Lupton and Miller, 1992). In Foucauldian terms the introduction of modern plumbing brought a new order to the private sphere without any need for direct forms of intervention to control social behaviour (Osborne, 1996). The role of water technologies within the home is testament to the sharpening gender differentiations and spatial occlusions associated with the emergence of the bacteriological city (Wright, 1975). Yet the modern bathroom has also become an 'outpost of freedom' for reflection and relaxation far removed from the communal baths of the past. The modern experience of water has effectively exchanged one form of bodily freedom for another as bathing has been transformed into an intimate exchange between privacy and running water (Busch, 1994: 68).

The bacteriological city reaches its zenith during the early decades of the twentieth century and comprises a series of identifiable features: the creation of large standardized technical networks driven by progressive conceptions of the role of science and technology in modern societies; the emerging role of infrastructure networks within the professional discourses of urban planning; the rise of the 'auto-house-electrical appliance-complex' as the focus for consumption within the modern city (Roobeek, 1987: 133); the centrality of infrastructure networks to 'the power, legitimacy and territorial definition of the modern nation state' (Graham and Marvin, 2001: 74); the ideal of universal access to sanitation as part of an integrative and standardized conception of environmental rights; and the creation of a new 'spatial fix' for capital investment in the built environment of the modern city (see Harvey, 1982; Lefebvre, 1991). During the early decades of the twentieth century the bacteriological model of scientific urban governance became fully developed as a managerial paradigm. The vast majority of homes in the cities of Europe and North America were now connected to centralized water supply systems. An acme of scientific urbanism was reached in which engineers could speak confidently of the complete control of urban metabolism as a technical ideal within their reach. The bacteriological city became allied with the advance of technological modernism through a progressive synthesis between water, space and technology. Yet at the moment of its triumph the bacteriological city and its vast infrastructural networks gradually disappear from view as part of the 'taken-for-granted' world of the modern city (see Kaïka and Swyngedouw, 2000). The routinization of spatial regulation contributes towards a gradual detachment of modern infrastructure from the legitimating ideals that made the political and administrative transformation of urban space historically possible (see Ingram, 1994: 253). We find that the scale of technological and institutional achievement is widely overlooked until the dramatic resurfacing of urban infrastructure in the 1970s and 1980s as the focus of a new urban problematic. The political and economic limits to the bacteriological city have now been exposed to intense scrutiny ranging from the neoliberal critique of public policy-making to the post-structuralist rejection of universalist ideals in urban planning. The physical fabric of the city has once again become a contested arena at the heart of urban political debate.

Urban infrastructure as neoliberal frontier

During the 1960s and 1970s historic associations between water, sanitation and progressive forms of urban governance began to fade. The role of capital in the determination of the physical and institutional configuration of urban space was enhanced in the face of the fiscal and ideological limits to the bacteriological mode of scientific urban management. The 'chilly limits' to engineering science initiated a new set of relationships between space, society and technology (Guillerme, 1988: 118). The supply-based engineering paradigm of universal water availability has been gradually displaced by a more differentiated conception of water as a commodity amenable to neoliberal modes of resource allocation. Water has become increasingly detached from the public realm as new kinds of technical, environmental and logistical demands have overwhelmed the capacity of under-resourced municipal authorities. New trends in water consumption such as the rapid growth of bottled drinking waters have served to weaken the ideological resonance of potable water as a shared good that is emblematic of an enlightened civic realm. Water itself has become increasingly commodified; or more accurately, we can argue that the relation between water and urban space has been appropriated by capital to a much greater extent than in the bacteriological city with a partial return to the private dominance of water supply that existed in the past.

The emergence of the 'fragmentary city' can be traced to several developments. The ideal of universal service provision has been undermined through a combination of pressures ranging from various forms of 'state failure' to the systematic dismantling of established patterns of collective consumption in the modern city. The removal of urban affairs from democratic control has been facilitated through immense privatization programmes involving the emergence of new corporate empires in fields such as water, energy and waste. These changes have precipitated a new wave of political conflict over urban infrastructure provision with consumer organizations, trade unions and anti-globalization activists pitted against a combination of corporate interests and neoliberal institutions (see Bakker, 2001, 2003; Swyngedouw, 1997, 1999). In Manila, for example, the much-vaunted water privatization of 1997 has run into serious political and economic difficulties. The devaluation of the Philippine peso has thrown the economics of privatization into disarray and led to escalating water tariffs. Foreign multinationals such as the French company Ondeo have reneged on earlier promises for city-wide service improvements and have focused their attention on more-affluent areas where residents can afford to pay higher tariffs (Chinai, 2002). Thus new forms of inequitable service provision have been superimposed over the uneven patterns of the past. The immense cost of modernizing dilapidated urban infrastructure has exposed the fragile capacity of municipal governments worldwide to manage their water and sanitation systems. The expense and complexity of new environmental discourses such as 'ecological modernization' have contributed towards the pressures to privatize state assets as governments seek to divest themselves of the escalating fiscal impact of infrastructure modernization.

Some indication of the current political hiatus is illustrated by the ill-fated Johannesburg environmental summit of 2002 in which the 'brown agenda' for improved global access to sanitation and drinking water became dominated by attempts to extend privatization and facilitate market access for global utilities. The international summits held during 2002 by GATT in Doha and the United Nations in Johannesburg reveal a growing coalescence of thinking between the Bretton Woods global financial institutions and the environmental and humanitarian programmes of the UN. A critical issue emerging is whether water supply and other municipal services can be subject to the same liberalization and free trade demands placed on others sectors of the global economy. These developments expose the scale of political and economic inequities that underlie international policy deliberations over issues such as inward investment, trade and debt. Immense political pressure has been placed on developing countries to open their markets for municipal service provision while the prospects for economic development are simultaneously thwarted by the inequitable terms of trade for primary commodities and manufactured goods. Attempts by municipal authorities in the South to emulate 'best practice' through the creation of locally-owned water utilities, for example, are stymied by their reliance on bilateral sources of capital rather than local sources of funding which provide better protection from the vagaries of international currency markets.

Though the fragmentary city shares some characteristics with its bacteriological, and even pre-bacteriological antecedents, we should be careful to acknowledge that contemporary urbanism is marked by a different set of relationships between capital, technology and urban form. Superficial comparisons between the nineteenth-century cities of Europe and North America, for example, and the contemporary mega cities of Asia, Africa and Latin America, overlook the political and economic dynamics of the imperial cities of the past, which were able to provide extensive investment opportunities for colonial capital. The combined effects of capital flight, debt and extremely rapid rates of growth, present a different dynamic for urban reconstruction in the cities of the South. We cannot invoke a technical genre of heroic engineering because Bazalgette, Haussmann and their contemporaries simply facilitated an urban transformation that was an outcome of the political and economic exigencies of their time.

More helpful than a historicist account of any putative urban transition is an attention to the emerging global dynamics of infrastructure investment patterns that bind disparate localities into a multi-tiered global arena. The emphasis of the bacteriological era on the universal allocation of basic services has been displaced by the development of more socially and spatially differentiated patterns of infrastructure provision. In the case of water this process extends to both production and consumption so that the denigration of the public realm in wealthy metropolitan regions has fostered a rapid increase in the consumption of bottled waters instead of 'untrustworthy' public supplies. This scenario of the rich partially disengaging from public water supplies provides an ironic contrast with the reliance of the urban poor on the exorbitant costs of private water vendors. The urban fringe of rapidly growing cities in the global South is marked by sometimes violent confrontations between municipal authorities and water vendor organizations who seek to prevent the extension of potable water networks into their areas of control. The most vulnerable sections of the urban poor thus find themselves trapped within highly inequitable networks of water provision fostered by the informalization of the urban economy and clientelist patterns of political power and patronage which militate against widespread improvements in urban infrastructure. The extensive 'splintering' of infrastructure networks, to use Graham and Marvin's term, is predicated on a much more fragmentary urban arena for capital investment in which social and spatial disparities become built into the physical fabric of the city. Whereas the extension of infrastructure under the aegis of the bacteriological city served to bind urban space into a more socially and spatially integrated system, the contemporary dynamic threatens an opposite process of intense social and spatial polarization.

Conclusions

Water provides a lens through which we can observe the complexities of urban politics ranging from the evolution of different forms of city governance to the micro-political domains of the home. The development of urban water systems reveals the centrality of public health as a focus for the

FIGURE 11.1 Andheri, Mumbai (2003). Where the bacteriological city intersects with the fragmentary city: giant water mains pass through urban slums with no access to piped water or sanitation. (Photo: Matthew Gandy)

understanding of both concrete processes of urbanization and changing conceptions of the urbanization process itself. There is an ambivalence that pervades the role of water as it oscillates between the discursive antinomies of the 'pathological city' and the 'progressive city'. In the modern era this historic dialectic between water and urbanism was extended to encompass a wider definition of the public realm in the face of new environmental challenges. Nineteenth-century approaches to the regulation of urban space sought detailed empirical information on every aspect of the modern city in a direct emulation of anatomical and histological advances within the medical sciences. A series of powerful circulatory metaphors emerged that informed the development of hygienist discourses in urban thought. Yet the bioregional ideal of the organic city conflicted with the social and economic dynamics of capitalist urbanization. The organic ideal of a synthesis between nature and culture, and between town and country, rested on a misconception of the commodification of nature within a modern capitalist economy. The hydrological utopia of the nineteenth-century hygienists went far beyond the political limits of laissez-faire urbanism as illustrated by the despair of Chadwick, Haussmann and other sanitary reformers over the waste of valuable human manure. The combination of changing attitudes towards human excrement, rising levels of water consumption, and new industrial sources of water pollution, made it impossible to secure an organic hydrological order.

The transition from the organic city to the bacteriological city involved a partial resolution to the contradictory dimensions of laissez-faire political economy. Questions of urban planning and design were brought within the domain of new administrative structures for the governance of cities. The diffusion of new water technologies and the spread of the modern bathroom allowed a rediscovery of the pleasures of bathing as part of a consumption-driven discourse rooted in the ideals of middle-class domesticity. The nineteenth-century tension between ascetic and sybaritic conceptions of water use was gradually resolved in the modern home through the spread of new consumption habits driven by the technological possibilities of modern plumbing systems. A 'water revolution' spread through modern societies predicated on the rapid diffusion of standardized technological networks. Yet in practice this universalist conception of water access remains historically fragile and geographically uneven. In colonial cities, for example, the universalist paradigm remained largely restricted to wealthy enclaves with only limited extensions made to other parts of the city. A distinction between 'citizens' and 'subjects' was translated into a brutal sanitary discourse of exclusion from the health benefits of the bacteriological city.

By the 1970s, however, even the limited conception of universal water provision integral to the bacteriological city began to lose its political legitimacy. The historic associations between water supply and emancipatory politics have been thrown into doubt by a radical diminution in the ideological coherence and political salience of the public realm. The contemporary drift towards the worldwide commodification of water has served to expose the political dimensions to urban infrastructure provision that had been temporarily obscured by the success of the bacteriological city and its centralized modes of technical intervention. The apparent consensus which has now emerged around the need for global improvements in urban sanitation masks fundamental political and economic tensions surrounding the scope and mechanisms of urban governance. The linkage of demands for water access with citizenship rights, for example, is suggestive of a very different kind of urbanism to the fragmentary consumer oriented models favoured by neoliberal policy makers. The danger is that the failures and limitations of the bacteriological model, which are now widely acknowledged, may be superseded by even less equitable responses to the water infrastructure crisis. Any emancipatory political agenda which lacks an adequate response to these questions has little hope of altering the material realities of everyday life for the urban poor.

Acknowledgments

I am grateful to Loretta Lees, Ben Page, and Rahul Srivastava for their detailed comments on an earlier draft of this chapter and to the UK Economic and Social Research Council for their financial support. Earlier versions of this chapter were presented to audiences in London and Mumbai during 2003 as part of an international seminar series entitled *Rethinking Urban Metabolism,* which brought together engineers, planners, and architects as well as a range of different academic disciplines, to explore the conceptual and practical dilemmas surrounding water, sanitation and urban infrastructure.

References

Armstrong, D. (1993) 'Public health spaces and the fabrication of identity', *Sociology,* 27: 393–410.

Bakker, K. (2001) 'From state to market?: water *mercantilización* in Spain', *Environment and Planning A* , 34: 767–90.

Bakker, K. (2003) 'Archipelagos and networks: urbanisation and water privatisation in the South', *The Geographical Journal,* 169(4): 328–41.

Benton, T. (1999) 'Scatology, eschatology, and the modern movement: urban planning and the facts of life', *Harvard Design Magazine,* Summer: 20–4.

Braham, W. W. (1997) 'Siegfried Giedion and the fascination of the tub', in N. Lahiji and D. S. Friedman (eds), *Plumbing: Sounding Modern Architecture.* New York: Princeton Architectural Press. pp. 201–24.

Busch, A. (1994) 'Geography of home, the bathroom: privacy and running water', *Metropolis,* 13(8): 68–9, 71.

Chadwick, E. (1885) *Royal Commission on Metropolitan Sewage Discharge, and the Combined and the Separate Systems of Town Drainage.* London: Longmans, Green & Co.

Chinai, R. (2002) 'Manila water system shows the failure of privatization', *Times of India,* 27 November.

Corbin, A. (1986) [1982] *The Foul and the Fragrant: Odour and the French Social Imagination.* Cambridge, MA: Harvard University Press.

Evans, R. J. (1987) *Death in Hamburg: Society and Politics in the Cholera Years 1830–1910.* Oxford: Oxford University Press.

Gandy, M. (1999) 'The Paris sewers and the rationalization of urban space', *Transactions of the Institute of British Geographers,* 24: 23–44.

Gandy, M. (2002) *Concrete and Clay: Reworking Nature in New York City.* Cambridge, MA: The MIT Press.

Giedion, S. (1948) *Mechanization Takes Command: A Contribution to Anonymous History.* New York: Oxford University Press.

Gordon, C. (1991) 'Governmental rationality: an introduction', in G. Burchell, C. Gordon, and P. Miller (eds), *The Foucault Effect: Studies in Governmentality.* Hemel Hempstead: Harvester Wheatsheaf. pp. 1–52.

Graham, S. and Marvin, S. (2001) *Splintering Urbanism: Networked Infrastructures, Technological Mobilities and the Urban Condition.* London: Routledge.

Guillerme, A. (1988) 'Sottosuolo e construzione della città/Underground and construction of the city', *Casabella: International Architectural Review,* 542/543: 118.

Harvey, D. (1982) *The Limits to Capital*. Oxford: Blackwell.

Ingram, D. (1994) 'Foucault and Habermas on the subject of reason', in G. Gutting (ed.), *The Cambridge Companion to Foucault*. Cambridge: Cambridge University Press. pp. 215–61.

Jacobson, C. D. and Tarr, J. A. (1994) 'The development of water works in the United States', *Rassegna: Themes in Architecture*, 57: 37–41.

Kaïka, M. and Swyngedouw, E. (2000) 'Fetishising the modern city: the phantasmagoria of urban technological networks', *International Journal of Urban and Regional Research*, 24: 120–38.

Klein, I. (1986) 'Urban development and death: Bombay City, 1870–1914', *Modern Asian Studies*, 20: 725–54.

Laporte, D. (2000) [1978] *History of Shit*, trans. N. Benabid and R. el-Khoury. Cambridge, MA: The MIT Press.

Lefebvre, H. (1991) [1974] *The Production of Space*, trans. D. Nicholson-Smith. Oxford: Blackwell.

Lupton, E. and Miller, J. A. (1992) *The Bathroom, the Kitchen, and the Aesthetics of Waste: A Process of Elimination*. New York: Princeton Architectural Press.

Melosi, M. V. (2000) *The Sanitary City: Urban Infrastructure from Colonial Times to the Present*. Baltimore, MD: John Hopkins University Press.

Osborne, T. (1996) 'Security and vitality: drains, liberalism and power in the nineteenth century', in A. Barry, T. Osborne and N. Rose (eds), *Foucault and Political Reason: Liberalism, Neoliberalism and the Rationalities of Government*. London: UCL Press. pp. 99–121.

Rabinow, P. (1982) 'Ordonnance, discipline, regulation: some reflections on urbanism', *Humanities in Society*, 5: 267–78.

Roobeek, A. J. (1987) 'The crisis of Fordism and the rise of a new technological paradigm', *Futures*, 19: 129–54.

Schorske, C. E. (1998) [1963] 'The idea of the city in European thought: Voltaire to Spengler', in C. E. Schorske, *Thinking with History: Explorations in the Passage to Modernism*. Princeton, NJ: Princeton University Press. pp. 37–55.

Sennett, R. (1994) *Flesh and Stone: The Body and the City in Western Civilization*. London: Faber and Faber.

Swyngedouw, E. (1997) 'Power, nature, and the city: the conquest of water and the political ecology of urbanization in Guayaquil, Ecuador: 1880–1990', *Environment and Planning A*, 29: 311–32.

Swyngedouw, E. (1999) 'Modernity and hybridity: nature, regenerationismo, and the production of the Spanish waterscape, 1890–1930', *Annals of the Association of American Geographers* 89: 443–65.

van Leeuwen, T. A. P. (1998) *The Springboard in the Pond: An Intimate History of the Swimming Pool*. Cambridge, MA: The MIT Press.

Vigarello, G. (1988) [1985] *Concepts of Cleanliness: Changing Attitudes in France since the Middle Ages*, trans. J. Birrell. Cambridge: Cambridge University Press.

Wright, G. (1975) 'Sweet and clean: the domestic landscape in the Progressive era', *Landscape* 20: 38–43.

12 In Search of the Horizon: utopia in *The Truman Show* and *The Matrix*

Geraldine Pratt and
Rose Marie San Juan

... you can picture the island as a sort of crescent, with its tips divided by a strait approximately eleven miles wide. Through this the sea flows in, and then spreads out into an enormous lake, though it really looks more like a vast standing pool, for, as it is completely protected from the wind by the surrounding land, the water never gets rough ... . The harbour mouth is alarmingly full of rocks and shoals. One of these presents no danger to shipping for it rises high out of the water ... [b]ut the other rocks are deadly, because you can't see them. Only the Utopians know where the safe channels are, so without a Utopian pilot it's practically impossible for a foreign ship to enter the harbour ... The town is surrounded by a thick, high wall, with towers and blockhouses at frequent intervals. On three sides of it there's also a moat, which contains no water, but is very broad and deep, and obstructed by a thorn-bush entanglement. On the fourth side the river serves as a moat. The streets are well designed ... The buildings are far from unimpressive, for they take the form of terraces, facing one another and running the whole length of the street. The fronts of the houses are separated by a twenty-foot carriageway. Behind them is a large garden, also as long as the street itself, and completely enclosed by the backs of other streets. Each house has a front door leading into the street, and a back door into the garden. In both cases they're double swing-doors, which open at a touch, and close automatically behind you. So anyone can go in and out – for there's no such thing as private property. (Thomas More, 1965: 69, 72–3, *Utopia*)

It is within a tradition of imagining utopic urban space that we locate the recent film, *The Truman Show*. The crescent shape of the island is distinct in its boundaries yet implies the presence of internal and external space. Its boundaries are porous to the outside yet entirely controlled from within. The internal environment calls up nature through the reassuring form of lush gardens, but nature's unpredictable forces are rarely in evidence. Urban space is strategically ordered yet distinctions between private and public space constantly disappear. Why do these attempts to produce a vision of urban perfection persist in evoking its edges?[1] Why are we led, in the film as in Thomas More's *Utopia*, to search for the limits of utopia's imaginative possibilities?

In Thomas More's *Utopia*, the island with its 54 identical cities is by its very definition a non-place. Such an imaginary construct provides a site from which to conceive a different society, a space for dialogue within an actual place, in this instance, More's own city of London. Only from such an imaginary site could a critique of social and political structures emerge, and as Louis Marin (1984: 3–30) has argued, it is in that dialogue, rather than in the utopic site itself, that new possibilities can be imagined.[2] And while this dialogue maps out space, and in particular an urban space, it is at the frontiers of that space that emancipatory opportunities seem to lie. Conceived as a horizon, this crucial point of passage implies both opportunities and unforeseen limits and costs. In other words, the productive aspects of utopic imaginings emerge at its horizon precisely because, as Marin (1993: 7–16) argues, the horizon simultaneously delimits a site and opens up a new space, providing 'a gap that does not allow any affirmation or negation to be asserted as a truth or as a falsehood'. More recently, Gilles Deleuze has also proposed the idea of a productive space of the imagination, an 'any space whatever' (1989: 156–88). This space cannot be described for it has no specified shape or boundaries but provokes thought through its ability to expose time disassociated from bodily action and narrative movement. For Deleuze, 'any space whatever' is a concept that emerges within a certain kind of film, a film that explores its ability to unhinge sequential time and, through discontinuous editing, disrupt any possibility of a place from which one is prompted to make normalizing moral judgements.

For all their differences, both More's *Utopia* and Deleuze's 'any space whatever' locate the possibility of new emancipatory thought within configurations of urban space. For both, a particular moment of urban crisis (Reformation London, European cities after World War II, respectively) presses the imagination to question the very forms of signification that sustain the status quo. Yet these are arguments that rest more with discursive modes of signification than with actual urban spaces, and for this reason can be profitably considered in relation to the work of Henri Lefebvre (1991), whose sustained critique of what he calls abstract space is particularly suggestive for thinking about the relation between utopic imaginings and actual social space. Lefebvre considers representational abstractions, and here one might include More's utopia, to be interlinked with capitalism's tendencies to abstract actual space as part of the process of producing and circulating capital. Urban space is ordered with the aim of erasing distinctions:

as much those which derive from nature and (historical) time as those which originate in the body (age, sex, ethnicity). The signification of this ensemble refers back to a sort of super-signification which escapes meaning's net: the functioning of capitalism, which contrives to be blatant and covert at one and the same time. (Lefebvre, 1991:49)

But abstract space harbours spatial contradictions, ones raised by the relation between actual and imagined space, or between what Lefebvre terms representations of space and lived time. For Lefebvre, as for Deleuze, time is a key factor in disrupting abstract space, especially historical time and aspects of lived time (for instance everyday life, death). These produce contradictions, especially within urban space, where places are simultaneously full of symbolic import and diminished by rational planning and organized networks of movement and information (Lefebvre, 1991: 231).

According to Lefebvre (1991: 385), then, there are very good reasons why we cannot give up on actual urban space as a site for new possibilities. In this chapter we take up Deleuze's invitation to use film as a vehicle for thinking but turn more directly to the predicament of the urban at a moment in which its emancipatory possibilities have been relocated within the imaginary spheres of technology. The city and technology have long been interlinked in both critique and emancipatory imaginings.[3] Recently, films such as *The Truman Show* and *The Matrix* have retained this crucial relationship and, at the very end of the millenium, have used it to explore and critique the concrete urban utopias that have emerged out of technology.[4] In this instance, the technology of film produces an imaginary non-place to critique itself, but one in which the status of the urban remains unresolved yet intricately connected to the critical possibilities of the relation between technology and the body.

Two utopias in question: new urbanism and the modernist city

In a doubled movement, *The Truman Show* and *The Matrix* present and critique utopic urban form. These films are particularly interesting to read in relation to each other because they are worrying about technology and the urban, but their worries are strikingly different, and they engage different urban forms: new urbanism and the modernist city, respectively. As a utopic site itself, film constructs the city through particular lenses, and the two utopic images of the city that emerge in these films are in part tied to attitudes towards technology. Yet, both films constantly move the viewer towards the edges of the utopic image, and their critical possibilities ultimately rest on what appears at the horizon.

The Truman Show is located in an actual utopic town, the celebrated Seaside in Florida, which is now taken by urbanists as the prototype of neo-traditional planning or 'new urbanism' (Falconer Al-Hindi and Till, 2001; Mohney and Easterling, 1991; Sexton, 1995).[5] Neotraditional planning has become an influential way of re-envisioning community in North American cities since the 1980s (McCann, 1995; Ross, 1999). It calls up a nostalgic pre-automobile town of white picket fences, architectural homogeneity, and small town sociability. In keeping with late nineteenth century 'garden city' ideals of merging town and country, every view of the town, renamed 'Seahaven' in the film, incorporates a garden landscape to suggest both the cultivation of nature and a detachment from urbanization. Nature is also evoked in the film through idealized sunsets and sunrises, with characters commenting on the setting's qualities of paradise. This dislocation outside dense urbanization is only heightened by scenes of bridges that stop abruptly just at the edge of the island. In the film the central character, Truman, himself embodies nostalgia, appearing in attire that is temporally ambivalent, somewhere between the present and the 1950s. His social exchanges likewise suggest a regularized, repetitive, non-conflictual way of life, as he moves from the domestic to public spaces that are all oddly intimate and familial. Even when Truman unexpectedly arrives at his wife's workplace in his pyjamas, this attracts no attention.

In fact Truman is an unsuspecting character in a television show, the only one who is unaware that he is living in a literal spectacle. He has been adopted by a television corporation at birth and has grown in front of an audience whose interest and delight in Truman's 'authenticity' recall current fascination with the 'real' and the way that technology makes it unfamiliar (e.g., living sites on the internet and reality TV). As Christof, the show's creator, puts it: 'While the world he inhabits is in some sense counterfeit, there is nothing fake about Truman himself ... it may not all be Shakespeare but it is real'. Thus the entire town is a stage set encased in a climate-controlled dome,[6] which, together with the Great Wall of China, is claimed to be one of two 'man-made' structures that can be seen from outer space.

Set up to raise the issue of viewing and its critique, the film is punctuated by scenes of two sets of viewers: the television show's audience, which is shown in semi-public and domestic places, and the show's orchestrators, who are shown in a high-tech control room located in the overseeing moon, and entirely encircled by multiple screens from which they choose views of Truman's daily activities. As Truman begins to question his world, the hidden viewpoints from which we see him become more overt (e.g., the vantage of a camera located in the lower dashboard of his car) and we become more concerned with issues of surveillance. After all, Truman is not only being watched, but his entire formation, including his memories, fears, and fantasies, are products of that spectacle. If he is afraid of the water, it is because the drowning death of his father, including Truman's own responsibility for it, had been scripted into the show. Ironically, this very fear has produced an increasing desire to leave Seahaven and explore places beyond. The control of Truman relies not only on his internalization of prescribed limits but also on an environment, replete with subliminal messages, that reinforces his fears. When he goes to the travel agent to acquire a plane ticket to Fiji, for example, one poster on the wall warns: 'Travellers Beware: Disease, Terrorism, Wild Animals, Street Gangs'; another, featuring a plane being struck by lightning, states 'This could happen to you!'

But how does Truman begin to recognize the conditions to which he has been subjected? There are two sources that are entirely in keeping with Lefebvre's notions of the potential ruptures of abstract space. The historical past, evacuated to keep the coherence of the present, returns. In *The Truman Show*, it returns in the form of the 'drowned' father when the disgruntled unemployed actor who played this role finds his way back onto the set, thus confronting Truman with an inconsistency in his story and his unresolved longings for his father. Rupture can also come from everyday movement. As Truman becomes more watchful, he detects difference in what was once the sameness of repetition. In a scene that has him wandering the street sensing the strangeness of his everyday world, he deviates from the usual pattern of walking and encounters what lies behind the back of the stage, namely actors taking a break and being prepared for the next scene. In another scene, sitting in his car on his driveway, he surreptitiously looks through his rear view mirror to detect the repetition of movement on the street, as the same Volkswagon 'Beetle', dog and bicyclist loop behind in his lane at five-minute intervals. Frustrated by his undefined suspicions, he attempts to escape the island by driving speedily and erratically, compelling those directing the show to quickly alter the usual traffic pattern. The third point of rupture does not comply with Lefebvre, for it is about a clear distinction between the abstraction and what lies behind it. Much of the humour of the film is in seeing these cracks in the staging, and invariably they are all attributed to glitches in the use of technology. For example, a camera light falls from the sky; or a sudden shower at the beach, intended to interrupt his melancholic ponderings on his situation, falls directly and only onto him.

In its conception of theatricality and its relationships of viewing, as well as in its sense of artifice that is distinct from the everyday, *The Truman Show* differs from *The Matrix*. Certainly the latter's conception of the utopic city is entirely different, opting for a modernist city of office towers, frequently viewed directly from above. Unlike the intimate public sphere of *The Truman Show*, this is a vision of corporate life inscribed through large corporate logos, literally written on the surface of the buildings. As Lefebvre characterizes abstract space:

> Differences, for their part, are forced into the symbolic forms of an art that is itself abstract. A symbolism derived from that mis-taking of sensory, sensual and sexual which is intrinsic to the things/signs of abstract space finds objective expression in derivative ways: monuments have a phallic aspect, towers exude arrogance, and the bureaucratic and political authoritarianism immanent to a repressive space is everywhere. (Lefebvre, 1991: 49)

Although filmed in Sydney, Australia, the views of the city in *The Matrix* are readjusted to house within it the history of technology (including the Eiffel tower), and, in a quintessential linkage between modernism and global technology, to stand for 'everycity'. It is not coincidental that most of the views are constructed by technology's ability to manipulate and distort (such as folding the city over the surface of a sphere or, in a representation that simulates video games, the fast movement of cars transformed into ribbons of colour and light). Considering this urbanism from a high viewing point, one of the characters states: 'Have you ever stood and stared at it, marvelled at its beauty, its genius? Millions of people just live out their lives, oblivious'. But there are also views closer to the ground and these have another sensibility, evoking the dark, mysterious, opaque spaces of 1930s noir film.[7]

What do we make of these contradictions? The world we are shown proves to be a computer program constructed to control human beings. We come to this realization together with Neo, the central character – a computer hacker by night, software engineer by day – who begins to receive perplexing messages that both recognize and heighten his ambiguous doubts about the veracity of his world. Even before he glimpses the truth, he asks a client purchasing some illegal entertainment software: 'You ever have that feeling where you don't know if you are awake or still dreaming?'. It is this type of questioning that leads him to be contacted by members of a renegade group, who are active in the dark, underside of the corporate city. Before they lead him out of the computer program to the real world, the film pinpoints what the costs might be through the explanation offered to Neo by Morpheus, the leader of the rebel group: 'The Matrix is everywhere, it is all around us, the world that has been pulled over your eyes to blind you from the truth: that you are a slave'. Ironically, Morpheus is the Roman god of dreams who can take human shape and induce mortals to escape worldly turmoil and enter a distracting dream (Hodgson, 2001: 1–5).

In *The Matrix* the encounter with the 'real' is the dream that propels Neo to leave the computer fantasy, but this is not to be a journey that can be reversed. Neo is asked to choose between a red and a blue pill: the blue leads him back to the computer fantasy with no memory of his doubts and suspicions, the red to a voyage of discovery to see 'how deep the rabbit hole goes'. He chooses the other side and awakes, foetus-like, in a container of transparent ooze, one of countless identical containers of imprisoned bodies stacked in a boundless chamber. He finds himself a different body: hairless, pallid, undeveloped muscles and organs, attached along his spinal cord to a vast machinery of electrical cords. Indeed, he has never lived outside of this cramped dish and, like all human beings, has been harvested to produce energy for the new superior mechanical beings. In an attempt to fight an all-out war against the takeover by machines, humans had created the climatic conditions that blocked the light of the sun. Now, in an unknown time in the future, all that remains of the city is miles of deteriorating underground sewers. The modernist city and the end of the twentieth century in which Neo believed he lived is just a mirage, a computer simulation into which he is hooked.

Why was Neo chosen to know this truth? Because chosen he was. Unlike *The Truman Show*, and more in line with Lefebvre, there is no easy way of stepping out of the world, for there is no behind to the screen. What Neo thought was his world is a sophisticated computer programme – 'The Matrix' – aimed to keep humans content in order to be productive. But even before he knows this, Neo dwells with some unease within the matrix. More than once Neo becomes aware of a discontinuity of what Lefebvre calls lived time, suddenly awakening when he presumed he was awake. According to Lefebvre, abstract space harbours contradictions that first become evident through the body's sensation of splintering time. Later in the film, the repetition of time, which Neo experiences as déjà vu, raises a glimpse of the system. When he sees the same black cat cross the same threshold, as if the same frame of film has been repeated, he is informed that it is a sign of a change in the system repairing itself. Thus, unlike *The Truman Show*, which suggests that in the hands of humans the presence of technology becomes evident through its glitches, in *The Matrix* the glitches in perception reveal the hidden struggle between technological masters and enslaved humans.

Actualized utopia

The films outline different personal costs in the attempt to actualize Utopia, costs that are posed at varying distances from the audience. In *The Truman Show*, living in a neotraditional utopia requires one to remain forever as a child contained within a web of fiction and frustrating doubts. This particular manifestation of abstract space is constructed within clichéd familial roles of child, parent, wife, and childhood friend, which as Lefebvre (1991: 49–50) suggests, purport to bring the bodily and sensual to the centre at the same time as they are abstracted. Jim Carrey, in the role of Truman, evokes an eerie combination of child, adult, and animated cartoon. In the most trying (and what should be intimate) moments, Truman's wife will insert a pitch for a domestic product, only accentuating the poignancy of his unfulfilled desires and the system's inept attempts to fulfill them. Of course this utopia is literally funded through such commercialization of everyday life and, as the director explains, involves not only 'product placement' but the commercialization of all that viewers see in the show. Conflict itself can only be addressed through similar means; at every moment of crisis, the traditional male 'buddy' shows up with a six-pack of beer. The beer as commodity is both advertising and an object that signifies and produces male sociability. Forced to live among actors scripted to be his wife, mother, father, or friend, Truman exists within a social environment of sameness that he yearns to leave. A teen romance with an actress, who is abruptly removed from the show when she attempts to tell him the truth of his situation, in part motivates this desire. But flashbacks to his childhood, for example when he declares to his teacher his desire to become an explorer, locate this yearning in the human condition. It is difference and the unexpected that Truman desires. When asked by his wife why he would want to go to Atlantic City, when he hates to gamble and the place would seem to hold no attraction for him, his simple reply is: 'I've never been there. Isn't that why people go places?'

And it is precisely conflict and the unexpected that engages the viewing audience, which is shown to take great pleasure whenever Truman transgresses the prescribed programme. This perception of authenticity leads us back to a lack on the part of the show's audiences, and, implicitly, the audience watching the film. After all, Truman's authenticity is only meaningful because the show is entirely orchestrated from above. The film gives to the creator of the show, Christof, the sole desire of maintaining the system without conflicts. Christof is portrayed as the creative mastermind, as the artist in a beret who dares to create in the place of God. His privileged eye holds control over a proliferation of technological eyes. When we are told that the surveillance of Truman has intensified from the initial camera that gazed on him in the womb to 5,000 cameras planted throughout idyllic Seahaven, the fears of the panopticon are fully in place.

A rather different dilemma is presented to the unsuspecting occupants of the Matrix. The harmonious, cohesive conception of a utopia critiqued in *The Truman Show* is already seen to have failed. We are told (by a machine) that when the machines originally devised the pacifying computer program in the mode of a traditional utopia, humans realized it all too well, saw the hands of power and refused to work: 'The first Matrix was designed to be a perfect human world where none suffered, where everyone would be happy. It was a disaster. No one would accept the progress. Crops were lost. As a species, human beings defined their reality from misery and suffering. The perfect world was a dream that your primitive cerebrum kept trying to wake up from ...' This resistance led the machines to revise the matrix into a system that incorporated within it struggle, conflict, and contradiction. They sited this utopia within the pinnacle of capitalist urbanization, with the desire to achieve and the motor of contradiction built right into it.

The film confronts us with the fact that we ourselves already may be within such a system. Even our belief in the materiality of the body, the place where we think there may be some residual matter, is constantly undermined. While critical theory has long been suspicious of visual perception in modern subjectivities (Crary, 1990; Jay, 1993; Silverman, 1996), other senses have been privileged as beyond corruption.[8] But in *The Matrix* this assumption is thrown into question: the rebels, facing a dull diet of porridge-like slop, contemplate how the Matrix programmed the appropriate food tastes: '[m]aybe they couldn't figure out what to make chicken taste like, which is why chicken tastes like everything'. In a counter scene, the Judas-like rebel, Cypher, at the moment of being wooed by one of the mechanical agents, launches into the pleasure of fantasy over a harsh reality. Holding a piece of steak, he questions the value of knowing reality if one believes the much more comfortable artifice to be real: 'I know this steak doesn't exist, I know that when I put it in my mouth, the Matrix is telling my brain that it is juicy and delicious. After nine years [of existing outside the Matrix] you know what I realized: ignorance is bliss'. He thus agrees to trade information that will destroy the rebels, on the condition that his memory will be erased and replaced. Thus memory also is shown, as in Cypher's worries of harbouring guilt through memory, to produce dissatisfaction, but also to be open for appropriation. When Neo returns to the Matrix and is moving through familiar urban spaces, he wistfully recognizes a favourite noodle shop, only to confront the artificiality of this construction, as of all his memories: 'I have these memories from my life; none of them happened'.

In sum, while *The Truman Show* offers some level of self-reflexivity, *The Matrix* brings critique that much closer to the body. In the former, we are positioned less as the spectacle than as audience, and we are given the opportunity to distance ourselves from the voyeuristic tendencies of those watching Truman, insofar as they are comically caricatured and thus positioned as another spectacle. In *The Matrix*, we are led to question that which we take as most basic: our bodies, senses, and memories. And the film is very clever in exploiting aspects of our perceptions that have already been brought into question, particularly as we become increasingly entangled with technology. After all, the taste of food, given changes in production and distribution, is already an issue in our societies (Ritzer, 2000). Neo's doubts may, after all, be our doubts.

In search of utopia's horizon

Both films, then, offer a critique of disembodied urban utopias, but it is our contention that – just like More's *Utopia* – it is that very site of critique produced between virtual and actual space that offers an opening to an emancipatory urban space. It is not incidental that for both films the most decisive site is the process of transition between the inside and outside. And as with Marin's (1993) notion of utopia's horizon, this point of rupture entails a recognition of the costs of change as well as its opportunities. But given that the films offer different critiques, they also differ in the ways they conceive of this transition. According to Deleuze (1989: 78–97), who is interested in the ways time criss-crosses and reveals the normative to be false, the virtual is always present in the actual and vice versa. If we have moved from the dangers of representation in *The Truman Show* to the site of the performative in *The*

Matrix,[9] we have also moved from a theatrical construct that we can escape and leave entirely behind, to a hegemonic construct that has no outside and is negotiated in the interrelation between mind and body.

It is telling that while for one film reaching the horizon brings the quest to completion, for the other it is just the beginning. After many failed attempts to escape, Truman conquers his fear of the water and uses a sailboat as a means of escape. In a scene that parallels the drowning of his father, Christof manufactures a ferocious storm to thwart his escape attempt. When Truman persists, Christof, to the disbelief of his production team, proceeds to undermine the utopic project and confront Truman with death. But Truman meets Christof's challenge and forges on only to encounter the actual limits of his world. As the effects of utopia have been spent and the environment begins to look more explicitly like a stage set, the boat punctures its paper-thin skin. Truman, in amazement, finally touches with his own hand the limits of his being, and then angrily bashes against its implications. Finding a stairway and at the top an exit button, he pushes it, opening a door. In a scenario that parallels Adam's departure from Eden, he is confronted with a momentous choice. The god-like voice of Christof is heard: 'I am the Creator ... of a television show ...' Truman: 'Who am I?' Christof: 'You are the star ... There is no more truth out there than there is in the world I created for you, same lies, same deceit but in my world you are safe. I know you better than you know yourself'. Truman: 'You never had a camera in my head'. While the door beckons, and in keeping with the theatricality that has informed the relationships of this world, Truman bids adieu to having been the spectacle. He reiterates his daily greeting to his neighbours: 'In case I don't see you, Good Afternoon, Good Evening and Good Night'; but now in a mocking way, and taking a flamboyant bow, he exits the stage. Thus Truman has recognized the necessary loss of childlike innocence and dependency, and accepted death as the ultimate cost of attaining selfhood. He crosses the threshold to enter the real city, and there is no going back because a clear line has been drawn between artifice and reality.

While the edge is crucial to *The Matrix*, it is but the beginning of the quest. Neo has to make the choice before the edge becomes visible and presents itself in the provocative form of the mirror, calling up images of *Alice Through the Looking Glass*, as well as Jean Cocteau's 1940 film noir, *Orphée*, but with a new twist.[10] Neo reaches with his hand to the mirror, only to have it liquify and encase him in a suffocating grasp. The film pulls all stops to convey the wonder and horror of a combined vision of technology and the human body as creators. For Neo still has to be born into something that we recognize as human, requiring initially an unplugging of his body from the technological system. Neo finds that he is soon tipped from the container of amniotic fluid-like liquid and sucked down a tunnel. He has been disposed of by the machines in order to be turned into feed for the human crop, but before this can happen he is snatched with a forceps-like instrument into the rebel ship. The tunnel thus functions both as a birth canal and a waste disposal. Christ-like, Neo is a being made muscle and flesh, which has within it simultaneously the processes of birth and death.

In Cocteau's *Orphée*, the other side of the mirror is the shadowy underworld where Orpheus is drawn in search of answers that can only be attained by confronting his own death. For Neo, the other side of the mirror entails both his growth from birth and resurrection from death. Displayed within a protective glass case, Neo is laid out in the manner of the dead Christ, punctured by numerous needles, tubes, and wires recalling the suffering wounds of Christ and medical procedures to bring forth the potential of the body (Kristeva, 1989: 110–18). As Lefebvre reminds us, abstractions of space only gain resonance because they are invariably intertwined with remnants of the past. In the case of *The Matrix*, the dazzle of technology gains its psychic grip through Christian iconography. The formation of Neo is accomplished through a whole range of technologies that extend him well beyond human limitations; sitting in a chair wired into the system, Neo's physical capabilities are increased exponentially through multiple computer programmes. What the other side of the mirror presents, then, is less an entirely different sphere, but the challenging relationship between the two.

While Morpheus believes that Neo is 'The One' who will free humanity from enslavement, the others have doubts, and none more so than Neo. Neo is put through various tests in the rebels' simulated matrix and, while deemed to be exceptional, he is made to face the limitations of his body. He fails, for example, the challenge to jump between corporate towers in a training computer program that simulates the modernist city of the Matrix. This is not unrelated to scenes in the rebel ship where Neo faces the very physical limitations of a decayed (actual) material reality. What the film is aiming for is a third option: the potential for something new emerging from the relationship between technology and material embodiment.

Faced with the question of whether Neo is 'The One', Morpheus takes him to consult the oracle. Neo is compliant but patronizing, and taking the position of the postmodern viewer, scoffs at the possibility of fate. The oracle is to be found in a run-down American high-rise, public housing project. Incorporating mixed signifiers, the oracle turns out to be a middle-aged, African American woman baking cookies in her kitchen. She treats Neo like a child, and sarcastically unsettles his system of beliefs. Within this highly technological world, once again we find the site of authentic wisdom in popular African American culture (Dyer, 1988). The oracle tells him: 'Sorry kid, you've got the gift, but it looks like you are waiting for somethin' … your next life maybe'. While Neo interprets her comments as a statement that he is not 'The One', the oracle presents him with a conundrum that he will have to face. The oracle thus offers Neo the means to further his potential, by showing him the road rather than giving him the answer.

Emancipatory possibilities

These films offer different emancipatory pathways. Truman must discover his inner strength – all that the camera cannot see. He learns to observe more closely, he arms himself (with a kitchen gadget that 'slices and dices', one of the many products marketed to the television audience by his 'wife'), and he masters his phobia of the water that has been cynically manufactured to keep him within the boundaries of Seahaven. He is an individual who must extract himself from the social relationships that entrap him, leave the artificiality of the constructed utopia, and conquer the natural elements (a violent storm at sea) in order to attain selfhood. The cliches of American citizenship – rugged, masculine individualism and mastery of nature at the frontier – are called upon and replayed. *The Matrix* also works with historical vestiges, in this case of folk culture and Christianity, to establish a path, but this path cannot be taken by the intact, self-sufficient individual. Neo's strength comes from using his human capacity for creativity and belief in relation, rather than in opposition, to technology. The boundaries of the body and subjectivity are utterly permeable, and this creates not only dangers but possibilities.

In neither film is emancipation located in a concrete place; in fact, emancipation is more about the capacity to leave particular spaces. Nonetheless Truman and Neo find their paths by means of and in relation to concrete urban sites. We might say that thinking concretely about emancipatory possibilities is not the same as concretizing utopic ideals. These films do the former but actively critique the latter. They allow us to think about emancipation through film and in relation to particular utopic non-places but they do not posit concrete alternatives. In *The Truman Show*, Seahaven is a site in which to contemplate the dangers of technological surveillance, desires for the architecturally-contrived safety of gated communities, and the pernicious effects of commodified culture. In an uncanny citation of Lefebvre's ideas about the ways that the urban and the lived experience of the body function as uncontrollable arenas of innovation and desire through which contradictions of abstract space emerge, Truman comes to his critical attitude through contradictions 'thrown up by historical time' (Lefebvre, 1991: 52), through differences produced 'unconsciously' out of repetitions' (1991: 395), and through literal tears in the fabric of sub/urban space. While Lefebvre argues that these disruptions exist in real cities, Seahaven offers additional critical possibilities precisely because it is a non-place against which to think about contemporary urban and anti-urban ideals.

But critical possibilities emerge in relation to particular imaginative geographies written into these non-places. In the case of Seahaven, the geography of liberalism is written into the filmic landscape, and it is thus not too surprising that the autonomous and self-determining individual is presented as our unlikely saviour from consumer capitalism. Michael Walzer (1984: 315) has defined liberalism as 'a certain way of drawing the map of the social and political world ... [Liberals practiced an] art of separation. They drew lines, marked off different realms, and created the sociopolitical map with which we are still familiar'. No line is more definitive than the line between public and private, and it is a line about which *The Truman Show* troubles a great deal. In a supposed interview at the beginning of the film, when the actress who plays Meryl, Truman's wife, is asked about the challenges of her role, she explains that 'for me, there is no difference between private and public life'. Yet the dissolution of this difference is at the heart of the film's critique of Seahaven. Truman has lost his privacy entirely.

If the public sphere within Seaside is solely artificial, little better can be said of the public sphere beyond it. The most public audience of the television show, to which we are returned repeatedly throughout the film, is located in a pub; the waitresses, bartender, and customers are all obsessively tuned to *The Truman Show,* and the only discussion among them turns around this televised spectacle. In his discussion of the historic decline of the bourgeois public sphere, Jürgen Habermas (1989) identified the eighteenth-century coffee house as an important space of vibrant public discussion. The pub in *The Truman Show* embodies what Habermas describes as the corruption of the public sphere, a society in which the one way transmission of televised, manipulated publicity has replaced intelligent public debate. The film's recovery of Truman's privacy as a means of rescuing a critical public sphere also resonates with Habermas's (1989: 157) location of its dissolution in the 'hollowing' of the family's intimate sphere. In Habermas's view, it was when the conjugal family lost its power as 'an agent of personal internalization which earlier facilitated the passage of universal values into the public sphere', that 'the source of light itself [for critical publicity] [was] now cast in shadow' (Cheah, 1995: 170).

The retrieval of political morality and public debate through the masculine individual and the revitalized conjugal family thus retraces and reverses Habermas's argument about the decline of the public sphere. Ironically, it is a strategy that repeats neotraditionalist values. To many this may seem a dubious and eminently conservative emancipatory strategy. Certainly the rights to mobility and freedom to explore differences that motivate Truman to wrest his subjecthood from his corporate parent hold within them the well-critiqued biases of liberal citizenship. Sherene Razack (2000: 95), for example, notes the traditional function of white men's adventures in urban space: 'moving from respectable space to degenerate space and back again is an adventure that serves to confirm that they are indeed white men in control who can survive a dangerous encounter with the racial Other and who have an unquestioned right to go anywhere and do anything'.[11]

But *The Truman Show* ought not to be discounted as a purely conservative film because it also raises an important debate about the collapse of public into private. The importance of doing this is suggested by Lauren

Berlant's (1997) discussions of infantile citizenship in the United States. She argues that the contemporary national imaginary in the United States is so rooted in traditional notions of family, home, and community that a non-conflictual, homogeneous space is taken as the ideal. In her view, citizenship has been 'downsized' to an act of voluntarism, and social justice has been replaced by a privatized ethic of responsibility, charity, and identification with suffering. One effect is that a conflictual public sphere has no place within this imagined nation: 'embodied activism performed in civic spaces has been designated as demonized, deranged, unclean (a) social activity' (1997: 179). Berlant (1997: 27) calls this 'infantile citizenship' because it manifests as passive patriotism and the suspension of critical judgement. This form of citizenship is something 'smaller than agency: [it is] patriotic inclination, default social membership, or the simple possession of a normal [and nor-malized] national character'. The intermingling of intimacy and citizenship has the effect, in Berlant's view, of impoverishing both. It is perhaps for this reason that an insistence on intimacy, as distinct from public life, is worth reassessing.

In *The Matrix* different critical possibilities are envisioned in relation to a different urban space. The geographies of liberalism have been erased by a fusion and confusion of public and private, inner and outer, dream and real-ity. In fact, the politics of liberalism are explicitly mocked. When Neo is cap-tured by the machines (which take the human form of the quintessential secret service agent – white skin, dark suit and sunglasses, short dark hair) and is taken into a small windowless room for interrogation about his con-tact with subversives, he cites his right to a telephone call. The agent finds this highly amusing and asks Neo about the utility of a phone if he has no mouth. The integrity of both the individual and social bodies are shown to be con-structed fictions as Neo's face contorts and his mouth is grown over by skin. Whereas Truman can taunt Christof with the knowledge that he 'never had a camera in [his] head', Neo comes to understand the total penetration of the Other within his mind and body, and the utter permeability of his being. Indeed, the ultimate sign of his power as 'The One' comes when he is able to penetrate the body of the machine/secret agent. Neo's body is not the intact liberal body and there is no reality entirely outside the Matrix and its technologies.

What are the emancipatory options within such a hegemonic system? At the end of the film, Neo walks the streets of the modernist city within the Matrix, hoping without certainty that he and the other rebels will be able to awaken other humans from their slumber. Nothing is in place. What he and the other rebels have created is an opening for awakening other humans to their condition and imagining another future that breaks the 'systematicity' of the matrix, what we might call abstract space. While the planned sequel to *The Matrix* may dispel our interpretation, we understand the first film to be positing human creativity, anti-systematicity, and persistent critique as an emancipatory model. Emancipation is a process rather than a form, although particular urban forms make visible the contradictions that engender this critical politics.

This reading betrays the influence of theorists such as Ernesto Laclau (1996) and Judith Butler (2000), who theorise the radicality and emancipatory potential of democratic politics and universal norms such as human rights in large part in their acceptance of peaceful social conflict as both a premise and virtue of public life. As Linda Zerrilli (1998: 15) phrases it, universality 'is not the container of a presence but the placeholder of an absence, not the substantive content but an empty place'. The emancipatory potential of democratic society, in other words, lies in an understanding that utopia as an actual place is unattainable – indeed undesirable. Emancipation is a never attainable process of critiquing the inevitable existence of social exclusions, rather than an urban or social form that attempts to resolve conflict. It involves a continual disruption of the abstractions within capitalist societies that work to mask social difference.

In making such an argument we are conscious and respectful of David Harvey's (2000) frustration with process-oriented utopias that ignore spatial materialization. In his view, these leave utopia 'as a pure signifier of hope destined never to acquire a material referent', with 'no way to define the port to which we might want to sail' (2000: 189). He urges the need for fantastic, concrete utopic visions to harness our imagination in order to counter the degenerative utopias of neoliberalism that presently envelop us. Indeed he has taken the risk of sketching such a place. But we want to suggest that, in light of current practices of infantile citizenship, a vision of utopia as a desired but unattainable horizon of possibility, as a concrete conflictual public sphere, is itself important, and we have argued, in a way that echoes rather than contradicts Harvey, that such conflicts come to be effectively articulated through and in relation to concrete urban forms.

In both *The Truman Show* and *The Matrix* the triumphant hero leaves the utopic city. Truman opens the door and steps out of Seahaven into the unknown. Neo flies up and out of the modernist city. The future is indeterminate – neither Seahaven nor the city beyond, the Matrix or the ruins of the modernist city contain or express the heroes' (or our own) desires. What they/we desire is something beyond abstract, instrumentalized space. All they offer is a horizon of possibility, a non-place in which to argue about concrete alternatives. But perhaps this is quite a lot.

Notes

1   This question of how the edge functions within critical assessments of urban life has also been raised in relation to the film, *Falling Down*. On how the edge and movement through an urban dystopia function in this film, see Morris (1999).

2 On the possibilities of utopia and diverse historical interpretations of this concept, see Schaer and Claeys (2000).

3 On the links between film, city and technology, see Easthope (1997), San Juan and Pratt (2002).

4 On the historical relation between utopia and millenial thinking, see Bann and Kumar (1993).

5 The architects of Seaside apparently proclaimed it as 'the second coming of the American small town' (quoted in McCann, 1995: 213). Critics of neo-traditionalists' social utopian claims to rebuild community across social diversity through careful urban design, argue that neotraditional planning is little more than a marketing tool that builds on negative images of the city and fears about the growing diversity of conventional suburbs. They argue that neotraditional developments are, in fact, socially exclusive (Dowling, 1998; Harvey, 2000; Till, 1993).

6 This dome calls up the original plans for Disney's EPCOT Center (Experimental Prototypical Community of Tomorrow) drawn up in the 1960s. EPCOT was to be an experimental new town, complete with 20,000 residents located within Walt Disney World and open to paying visitors. It was designed as a radial city in which the downtown core of 50 acres was covered by a glass dome to keep out inclement weather. Skyscrapers would pierce the dome but smaller buildings, including housing, stores, offices, theatres and nightclubs, would be contained within the climate controlling dome (Warren, 1993).

7 This tension between the modernist city and its dark underside is a recurring one in films of the modern city. See Donald (1999), Easthope (1997), Straw (1997).

8 The Cartesian mistrust of vision, which remains embedded in twentieth century French critical theory, has led to a distinction between sight (external) and the other senses (internal) in relation to the body. On the potential of other senses to retain a closer link to the materiality of the body, see Corbin, (1994). On haptic visuality that draws on Deleuze's notion of heightened senses, see Marks (2000). For a critique of this divide, see Bal (1999: 234–5).

9 By using the term performative we are invoking Butler's (1990) notion that there is no distance between social norms and the interior spaces of the subject.

10 The appearance of the rebels in *The Matrix*, defined by stylish militaristic costume and weaponry, bears a striking resemblance to Cocteau's underworld characters, which convey danger by drawing on the memory of recent French struggles with the war-time fascistic government.

11 This echoes hooks's (1998) influential discussion of the politics of cosmopolitan consumption of 'the other' and white practices of 'eating the other', whether it be through touristic sexual adventures or eating in 'authentic' ethnic restaurants. She argues that the impulse behind this touristic cosmopolitanism is self-transformation. The Other is seen as a source of life-sustaining alternatives to a drab, exhausted white culture but such excursions into otherness are embarked upon with the understanding that hegemonic white culture will remain unchanged. The goal is self-transformation and not transformation of one's world.

References

Bal, M. (1999) *Quoting Caravaggio: Contemporary Art, Preposterous History.* Chicago, IL: University of Chicago Press.

Bann S. and Kumar, K. (eds) (1993) *Utopias and the Millennium.* London: Reaktion Books.

Berlant, L. (1997) *The Queen of America Goes to Washington City.* Durham, NC: Duke University Press.

Butler, J. (1990) *Gender Trouble: Feminism and the Subversion of Identity.* New York: Routledge.

Butler, J. (2000) 'Restaging the universal: hegemony and the limits of formalism', in J. Butler, E. Laclau and S. Zizek (eds), *Contingency, Hegemony, Universality: Contemporary Dialogues on the Left.* London: Verso. pp. 11–43.

Cheah, P. (1995) 'Violent Light: the idea of publicness in modern philosophy and in global neocolonialism', *Social Text*, 43: 163–90.

Corbin, A. (1994) *The Foul and the Fragrant: Odour and the Social Imagination*, trans. M. Koshan. London: Picador.

Crary, J. (1990) *Techniques of the Observer: On Vision and Modernity in the Nineteenth Century.* Cambridge, MA: MIT Press.

Deleuze, G. (1989) *Cinema 2. The Time Image*, trans. H. Tomlinson and R. Galeta. London: Athlone Press.

Donald, J. (1999) *Imagining the Modern City.* Minneapolis, MN: University of Minnesota Press.

Dowling, R. (1998) 'Neotraditionalism in the suburban landscape: cultural geographies of exclusion in Vancouver, Canada', *Urban Geography*, 19: 105–22.

Dyer, R. (1988) 'White', *Screen*, 29: 44–65.

Easthope, A. (1997) 'Cinecities in the Sixties', in D. B. Clarke (ed.), *The 'Cinematic' City.* London: Routledge. pp. 129–39.

Falconer Al-Hindi, K. and Till, K. (2001) '(Re)placing the new urbanism debates: toward an interdisciplinary research agenda', *Urban Geography*, 22: 189–201.

Habermas, J. (1989) *The Structural Transformation of the Public Sphere*, trans. T. Burger. Cambridge, MA: The MIT Press.

Harvey, D. (2000) *Spaces of Hope.* Edinburgh: Edinburgh University Press.

Hodgson, B. (2001) *In the Arms of Morpheus: The Tragic History of Laudanum, Morphine and Patent Medicines.* Vancouver: Grey Stone Books.

hooks, b. (1998) 'Eating the other: desire and resistance', in R. Scapp and B. Seitz (eds), *Eating Culture.* Albany, NY: SUNY. pp. 181–200.

Jay, M. (1993) *Downcast Eyes: The Denigration of Vision in Twentieth-Century French Thought.* Berkeley, CA: University of California Press.

Kristeva, J. (1989) 'Holbein's dead Christ', in J. Kristeva, *The Black Sun: Depression and Melancholia*, trans. L. S. Roudiez. New York: Columbia University Press. pp. 105–38.

Laclau, E. (1996) *Emancipation(s).* London: Verso.

Lefebvre, H. (1991) *The Production of Space*, trans. D. Nicholson-Smith. Oxford: Blackwell.

Marin, L. (1984) *Utopics: Spatial Plays*, trans. R. Vollrath. Atlantic City, NJ: Humanities Press.

Marin, L. (1993) 'The frontiers of utopia', in S. Bann and K. Kumar (eds), *Utopias and the Millennium*. London: Reaktion Books. pp. 5–16.

Marks, L. (2000) *The Skin of the Film: Intercultural Cinema, Embodiment, and the Senses*. Durham, NC: Duke University Press.

McCann, E. (1995) 'Neotraditional developments: the anatomy of a new urban form', *Urban Geography*, 16: 210–33.

Mohney, D. and Easterling, K. (eds) (1991) *Seaside: Making a Town in America*. New York: Princeton Architectural Press.

More, T. (1965) *Utopia* [1521] trans. P. Turner. Harmondsworth: Penguin Books.

Morris, B. (1999) 'Warzones of the street', in L. Finch and C. McConville (eds) *Gritty Cities: Images of the Urban*. Sydney, Australia: Pluto Press. pp. 147–61.

Razack, S. (2000) 'Gendered racial violence and spatialized justice: the murder of Pamela George', *Canadian Journal of Law and Society*, 15: 91–130.

Ritzer, G. (2000) *McDonaldization of Society*. Thousand Oaks, CA: Pine Forge Press.

Ross, A. (1999) *The Celebration Chronicles: Life, Liberty and the Pursuit of Property Values in Disney's New Town*. New York: Ballentine Books.

San Juan, R. M. and Pratt, G. (2002) 'Virtual cities: film and the urban mapping of virtual space', *Screen*, 43: 250–70.

Schaer, R. and Claeys, G. (eds) (2000) *Utopia: The Search for the Ideal Society in the Western World*. Oxford: Oxford University Press.

Sexton, R. (1995) *Parallel Utopias: The Quest for Community: The Sea Ranch, California, Seaside, Florida*. San Francisco, CA: Chronicle Books.

Silverman, K. (1996) *The Threshold of the Visible World*. London: Routledge.

Straw, W. (1997) 'Urban confidential: the lurid City of the 1950's', in D. B. Clarke (ed.), *The 'Cinematic' City*. London: Routledge. pp. 110–28.

Till, K. (1993) 'Neotraditional towns and urban villages: the cultural production of a geography of otherness', *Environment and Planning D: Society and Space*, 11: 709–32.

Walzer, M. (1984) 'Liberalism and the art of separation', *Political Theory*, 12: 315–30.

Warren, S. (1993) 'The city as theme park and the theme park as city: amusement space, urban form and cultural change'. Unpublished PhD. Department of Geography, University of British Columbia.

Zerrilli, L. M. G. (1998) 'This universalism which is not one', *Diacritics*, 28: 3–20.

13 Ghosts and the City of Hope

Steve Pile

The tradition of the dead generations weighs like a nightmare on the minds of the living. (Karl Marx, 1852 [1973]: 146, *The Eighteenth Brumaire of Louis Bonaparte*)

The social revolution of the nineteenth century can only create its poetry from the future, not from the past. It cannot begin its own work until it has sloughed off all its superstitious regard for the past. Earlier revolutions have needed world-historical reminiscences to deaden their awareness of their own content. In order to arrive at its own content the revolution of the nineteenth century must let the dead bury their dead. (Karl Marx, 1852 [1973]: 146, *The Eighteenth Brumaire of Louis Bonaparte*)

It is necessary to speak *of the* ghost, indeed *to the* ghost and *with* it, from the moment that no ethics, no politics, whether revolutionary or not, seems possible and thinkable and *just* that does not recognize in its principle the respect for those others who are no longer or for those others who are not yet *there*, presently living, whether they are already dead or not yet born. (Jacques Derrida, 1993: xix, *Spectres of Marx*)

Politics must contend with the dead, with ghosts – well, according to Marx and Derrida, that is. If we're to believe them, then any story we might wish to tell about the emancipatory city would have to take into account its ghosts. On the one side, Marx would be there telling us that we have to let go of the spirits of the past; we revolutionaries must let the dead bury the dead. On the other side, Derrida tells us that it is not so easy to exorcise our ghosts; instead, in the name of justice, for those who have died, who have not yet died, and who have not yet been born, we must speak to and with the ghost. If the city is to be emancipatory, then, there is a troubling problem: what is to be done with the tradition of dead generations that weighs so heavily on the lives of city dwellers? Perhaps there is a prior question: do traditions haunt the city, like ghosts? Let us see: for events in London in 2000 seemed to suggest they do.

Grave Situations: London, 1 May 2000

Politics are an ever-present aspect of city life. In other words, cities are intense sites of political activity. London – so-called global city, seat of the British government, once heart of Empire – is no exception. In this section I will describe some of the events surrounding the anti-capitalism protests around Parliament Square and Trafalgar Square on 1 May 2000 (unofficially Labour Day in the UK). Here, I will be seeking to draw out the emotional intensity of these protests by showing how ghosts came to haunt certain actions. Of course, ghosts haunt cities in many ways – including, as we will see, the spectres of a future yet to come. In the section that follows, I will explore the liberatory possibilities that disparate and diverse oppositional movements offer the city through a discussion of David Harvey's dream of a utopian future (2000: 257–81). Throughout this chapter, my focus is on the 'other-worlds' that connect the emotional geographies of protest with the political geographies of emancipation. Here, the social figure of the ghost evokes both the nature of these other-worldly connections and their (political) effects. Moreover, the ghost will tell us about the discontinuities of time and space that are present (and absent) in the production of geographies of the city. Let me begin with the events of 1 May 2000.

On 1 May 2000, a crowd, estimated at 4,000 people, gathered in London's Parliament Square to protest against the ills of contemporary capitalism (The *Guardian*, 2000: front page). For many people, this was a further example of an increasingly radical (and violent) turn in popular protest. It bore echoes of rioting that had occurred the previous year in London's financial district on 18 June, as well as events in other cities around the world, such as (most famously) Seattle (see St. Clair, 1999). The police, not wishing to be caught unprepared (as they had been at the J18 demonstration), had mobilized the biggest operation in the capital for 30 years.[1] There were 5,500 officers at the event, with a further 9,000 officers in reserve: a total, then, of about 14,500 officers.[2] Even acknowledging the likely undercounting of the number of protesters (one would normally double official estimates), this would make as many police as protesters. Let us think about this: so, a bigger police operation than during the anti-nuclear, anti-war, trades union, feminist, anti-racism and anti-apartheid demonstrations of the 1970s and 1980s (and 1990s). Somehow, the brief, but explosive, violence of the J18 protests had put the police – and the authorities – on their guard. What revolutionaries were these?

Protesters had been gathering around Parliament Square from about 10.00am onwards. Perhaps unsurprisingly, several hundred cyclists arrived first, then anarchists and communists, situationists, ecologists and gardeners, and even a few tourists. Though the main 'disorganization' at the heart of the demonstration was *Reclaim the Streets* (see Gavin Brown, this volume), the diversity of groups was visible in the many flags and banners that festooned the streets. But particular prominence was paid, in the press, to the direct action known as 'guerrilla gardening': some described it as a typical English garden party, but without the tea!

By 11.00am, green-fingered activists were busy digging up the grassy area outside parliament and turning it, rapidly, into a garden – planting seeds, vegetables and flowers – complete with ponds and the odd gnome. The turf from the diggings was taken and laid on the roads. Squares, roads and streets were being reclaimed. Even the smell of the city had changed: from that of carbon monoxide and tarmac to that of earth, turf and manure. There was music too: a samba band played, drummers went a-drumming, and there was dancing in the streets. All a bit untidy, a bit chaotic, and perhaps a bit extreme or nonsensical, but somehow not that bad.

'Not that bad' was about to turn bad, for there was trouble in store. Sometime around 1.40pm, some of the protesters began to attack various businesses in Whitehall, on the fringe of Trafalgar Square. A couple of young men were caught on CCTV breaking the front window of McDonalds; one wearing a black mask, gloves, and wielding a chair (and a carrier-bag with what looked like two Coca-Cola bottles in it) (see Figure 13.1). Soon, the

FIGURE 13.1   The McDonalds in Whitehall is attacked by protesters, 1 May 2000 (Reproduced by kind permission of PA Photos)

signature M sign was pulled down and destroyed; 'McExploit' had been sprayed in red onto the wall; 'McShit' in black. Next door, a *bureau de change* was broken into, as was a nearby souvenir store (?!). Guerrilla gardening was turning by degrees into a set of attacks on properties that signified, in some way, all the evils of capitalism (in its global guises).

Now property was under threat, and the police intervened. At around 2.30pm, the police acted in defence of law, order and, it must be added, capitalism in all its guises. In full paramilitary paraphernalia, baton-wielding, body-armoured ranks of police began smashing protesters' unarmed, unprotected bodies. Whether window-smashing anarchists or plant-wielding gardeners hardly mattered. It's not as if the police could distinguish between them anyway. Guerrilla gardening quickly became guerrilla warfare, with short battles being fought around Trafalgar Square for almost 4 hours. From the beginning, the police had blocked all possible escape routes and begun to squash the protesters into Trafalgar Square. By 6pm, the police (in their words) retook Trafalgar Square. Though sporadic 'rioting' continued into the evening (as the police pushed small groups of demonstrators up the Strand), civilization and the city at last were saved and safe. Now was time for political comment. Prime Minister Tony Blair weighed in with this:

The people responsible for the damage in London today are an absolute disgrace. Their actions have got nothing to do with convictions or beliefs and everything to do with mindless thuggery.[3] (quoted in the *Guardian* 2000: 1)

Let us be clear, Tony Blair was not referring to the police in his remarks. The police, nevertheless, became subject to bitter criticism – because of their own version of mindless thuggery. With their overwhelming force, it was wondered why they hadn't intervened earlier. It was also questioned why they had funnelled all protesters into Trafalgar Square, rather than allowing peaceful demonstrators to escape. However, criticism of the police was ultimately muted. Indeed, the police had come under greater criticism after the J18 riots for not being prepared for, and not using, full force.

Predictably, there was a failure on the part of parliamentarians to see the grace and mindfulness of the protests. And not just politicians, liberal political commentators bemoaned the lack of a (decent) political agenda in the demonstrations: heavy-weight journalist Hugo Young (2000: 18) interpreted the actions as stemming from 'a bogus romance with anti-politics'. Hugo Young even underscored the futility of the 1 May protests by making unfavourable comparisons with the demonstrations in Seattle and Washington, which he argued had intensified political debate on the consequences of global capitalism – to the benefit of consumers!

Now, such controversies would probably have arisen over any such event in London. Few had said good things about the Poll Tax riots ten years earlier. Nevertheless, these red and green revolutionaries were increasingly being hidden by legions of ghosts, as they marched right through the May Day protests. Out of the wreckage of this little part of London, hundreds of thousands of the dead were now haunting the minds of the living; these spirits, however, were resolutely on the side of the authorities – for, what else had they died for but 'us'? For, there are ghosts in cities.[4] Let us meet them (even if we cannot yet talk *to* them, as Derrida would wish).

Ghost number 1: Churchill. There is a statue of Winston Churchill in Parliament Square. Early on in the day, the stern, glowering dark metal effigy had attracted the attention of activists. A dash of red paint made 'blood' stream from the corner of his mouth. In his right hand, a police officer's helmet hung by its strap, and in the helmet rested a geranium; other symbols, in red, appeared, while activists carrying banners stood on the plinth, two carrying a large yellow banner with TIKB[5] in large red letters, all with happier expressions than Winston. Throughout the day, Winston's own mood seemed to darken, until finally someone added a strip of turf to his bald head, thereby giving Winston a green 'Mohican' (see Figure 13.2).

FIGURE 13.2 Churchill defaced, 1 May 2000 (Reproduced by kind permission of PA Photos)

It was this image that stared out from the front pages of the popular press, but not to make people laugh or raise people's consciousness about the environmental consequences of global capitalism. Instead, Winston glowered, indignant. For the desecration of the statue was also a desecration of all that Winston Churchill (literally) stood for: the blood, sweat and tears of the Second World War, when Britain stood alone (it is said) for democracy, freedom, and civilization against the forces of fascism, slavery, and barbarism. Standing alone, now, Winston was covered in paint, dirt and grass; his ghost forcing the memory of that sacrifice, but also forcing other (past) political visions into the foreground.[6] Older visions began to cast a dark shadow over the still shimmering political visions of the protesters. The tradition of that dead generation weighed heavily on the attempts of the living to revolutionize traditions of democracy, freedom, and justice. Suddenly, an alternative flow of history was being dammed: the past stood firm, sullenly accusing the present of its failure to live up to the dreams of the past. Winston's ghost, however, was not alone: hundreds of thousands of others joined him – far angrier, it would seem, than he.

Sometime between 2pm and 4pm, probably, the Cenotaph in Whitehall had been defaced. The Cenotaph in Whitehall is Britain's monument to its war dead. Primarily, it is associated with the dead of the two world wars – 1914–1918 and 1939–1945 – although it is actually intended to commemorate all of 'the fallen', including those up to the present day. Every year, on a Sunday in early November, Remembrance Day is observed by royalty, political leaders, the armed services, veterans and relatives, when they gather at the Cenotaph to lay wreaths and pay their respects to the dead. A minute's silence is observed, then broken by buglers sounding the Last Post. Instead of samba and dancing, memories of Rule Britannia echo through London's deadened streets. The monument itself is made simply of white stone, with a kind of stark presence that itself calls to mind the grim finality of death (see Figure 13.3).

For many, it was the desecration of this national symbol of sacrifice and grieving that most starkly told of the 'mindlessness' of the demonstrators. Most offensive of the graffiti was this: on one side of the war memorial two green arrows, one marked 'MENS' and the other 'WOMENS', indicated where the 'TOILETS' were. The ultimate sacrifice of hundreds of thousands of (mainly) young men and (many) women had been profaned. On the evening's television news programmes, survivors of both world wars and relatives of the fallen vehemently and poignantly expressed their utter contempt for the protesters and, by implication, the protests – their words (as you might anticipate) juxtaposed with familiar, yet ghostly, images of youthful soldiers going 'over the top' in the First World War. Instead of the dead burying the dead, the dead were burying the living under the cold earth of an unrealized future; the future that those men and women had died for. The ghosts altered reality: now, the protests were haunted by an idealized past and an idealized

FIGURE 13.3   The Cenotaph as it usually looks. The inscription reads 'The Glorious Dead' (Photo: Steve Pile)

(but unreal) future. Shudders ran down everyone's spine. These ghosts are impossible to ignore, *especially* when they are being ill-treated.

It is reasonable (and fairly easy) to argue, counter all this, that there are no ghosts at play in these situations. But this is to misunderstand the nature of ghosts and their relationship to cities, and even the spectrality of cities. Ghosts take on many guises in the story above: we have, of course, Churchill's admonishing body; large armies of men, in uniform, covered in blood; we have the spectres of dead ideas – revolutionary or not, vegetarian or not, militaristic or not, democratic or not. There is even the ghost of

England's green and pleasant land and of 'Jerusalem', which London has so evidently failed to become. These spectres call forth ideas, and also feelings. The indignity and injustice of death returns (once again) to haunt the living. And the living are (once again) caught up in the traumas and losses of the past. As Walter Benjamin (1999 [1935]) might observe, at this point we can see various 'mechanisms' through which 'the dialectic' of history is brought to a shocking stand-still as the dead cling like chewing-gum to the heels of the living. In part, the imaginations of the living call to life the dead generations as they shimmer in photographs and old newsreels or in the testimonies of those who remember. We can take this argument a step further: for the dead also hold fast to the living through the spectral geographies of the city.

Ghosts haunt; to haunt is to possess some place. There are ghosts 'in place': in this instance, the ghosts of Britain's dead. There are ghosts 'out of place': spectres, for example, of Marx. These are the collective ghosts of selective and partial memory – suffering, again, from their exhumation in the name of truth and justice. Such resurrections can, moreover, confirm the partiality of the dead, as they circulate among the living. Among the demonstrators, the ghosts of different radical traditions faced off against one another: as anarchists burnt books by Marx, reasoning that he was an authority figure; as the odd communist or anarchist took time off from protesting against capitalist fat cats and pigs to shout 'Fucking Hippies' at the guerrilla gardeners. The living don't always let the dead remain buried – and they don't always respect them once they've dug them up.

There is more to say, of course, but the reason for telling these stories in this way is to get at *something* to do with the ways that cities – and not just those cities that are present in the place we call London – contain their pasts and how these pasts take hold of the present and the future. This 'something' is a fractured emotional geography cut across by the shards of pain, loss, injustice, and failure: an emotional world in which the ghost is the emblematic resident. More than this, the streets of London – just as the ghosts of other cities – take hold of the imagination: they defy us to find hope, since they have none; they defy us to make reparation, since none can be found for them; they defy us to allow them to bury themselves, since we have already buried them; they defy us to find new traditions, since we can never atone for the fact that this means killing them once again – this time, stone dead.

Ghosts, we can say, haunt the places where cities are out of joint; out of joint in terms of both time and space.[7] They join the living world through fissures. They grip the imagination, screaming at the trauma and pain they have had to endure, lamenting the injustices that have befallen them. Of course, anyone with any sense of justice would have to listen to tales of loss and torment – which is why we will turn next to David Harvey, whose sense of the hidden injuries of class is acute. Such tales of pain and injustice are common-place in cities, but our relation to these stories ultimately determines how (and whether) we can revolutionize the present – and, therefore, the future. What grip, then, do ghosts have on our stories of the city of hope?

Hope and the haunted city: utopian dreams, dystopian nightmares

The phrase 'city of hope' in the title of this chapter is an echo of David Harvey's (2000) book, *Spaces of Hope*. In this book, David Harvey wishes to set out the resources of hope for thinking about a better, even utopian, future. Harvey's own title, it should be added, is itself a reference to Raymond Williams's (1989) 'last' book, *Resources of Hope*. And, thereby, situates itself firmly in a long line of utopian thinking about cities, stretching back at least as far as Ebenezer Howard's (1902) *Garden Cities of Tomorrow*. However, Harvey refuses to allow himself to be limited to debates either within urban planning or those among contemporary urban critics. He also draws on a strain of imaginative utopian writings by novelists and science fiction writers, linking his thought most explicitly with Edward Bellamy's (2001) utopian fantasy *Looking Backward, 2000–1887*. In some ways, then, Harvey's dialectical imaginings for the city involve a comparison between the present as it actually is and the present as it might have become (had the Victorians actually set us on the path to build the Jerusalem they dreamt of). The ghost of dead ideal cities shimmer in the present, disturbing the solidity of living generations (and adding to their nightmares). Harvey's wish, like those of the May Day protesters, is to revolutionise history: to dream anew the old dreams of utopia. Such dreaming is instructive for cities that long for, or wish to be sites of, emancipation.

Let us start, then, with Harvey's judgement about the current relationship between (older) utopian longings and the condition of the city. With the passing of the high hopes of the revolutionaries and reformers of the nineteenth century, Harvey surmizes, much – perhaps everything – has been lost:

> Utopian longing has given way to unemployment, discrimination, despair, and alienation. Repressions and anger are now everywhere in evidence. There is no intellectual or aesthetic defence against them. Signs don't even matter in any fundamental sense any more. The city incarcerates the underprivileged and further marginalizes them in relation to broader society. (2000: 11)

To substantiate this broad claim, Harvey spends time looking in great detail at the condition of Baltimore: a city he had already examined in *Social Justice and the City* (1973). Looking backward from 2000 to 1973, things could hardly be said to have improved: and they weren't exactly great in

1973! In Harvey's judgement 'Baltimore is, for the most part, a mess. Not the kind of enchanting mess that makes cities such interesting places to explore, but an awful mess' (2000: 133). The extent of this mess is dispiriting: it is evidenced in vacant houses, homelessness, unemployment, the percentage of people receiving welfare payments, soup kitchens, charity missions, inequality, poor employment (involving the replacement of relatively well paid, full-time industrial employment by low paid, part-time and temporary service jobs), poor schools, low educational attainment, social distress, poor health, suburban sprawl (with its conformity and environmental unfriendliness), deindustrialization, and a monstrous downtown.

The renewal of Baltimore's downtown does seem to have a good side, with the Inner Harbor attracting a mix and mixing of people, creating a vibrant 24-hour city. However, for Harvey (2000: 141), such vibrancy comes at an unseen cost: it is underpinned by a 'hefty dose of social control' and gentrification. Thus, the enjoyable aspects of Baltimore's new social mixing turn out to be superficial and disingenuous, for these developments never touch 'the roots of Baltimore's problems' (2000: 148). Worse, the renewal programmes for the downtown area and the suburbs have actually sucked much needed public funds away from the inner cities and, therefore, away from those most in need of public support. Uneven geographical development is thereby intensified by the intervention of the state, which intensifies the already unequal and uneven tendencies within capitalist investment and dis-investment strategies (2000: 150–154). Baltimore, in Harvey's view, is not a special case, but almost paradigmatic of urban development across the United States and, by extension, the rest of the liberal democratic capitalist world.

In such circumstances, the spaces of hope are hard to find; for the most part, hope seems to have died – in ways that are reminiscent of James Thomson's (1880) bitter poem about the dreadful state of Victorian cities. If Faith, Hope, and Charity died in the Victorian city, in Baltimore, more promisingly, all three seem to be struggling on together in the soup kitchens and church missions. Nevertheless, utopian longings seem a luxury. Whether too despairing in the gutted hand-to-mouth inner city, or too taken with the glittering attractions of the soulless middle-class entrepreneurial city, few seem willing to play with utopian ideas or to think about how to convert utopian ideas into future realities. Despite the best efforts of Faith, Hope, and Charity, Utopia seems to have died a death.

Harvey's analysis of the death of utopian longing reveals two causes, each the opposite of the other.[8] On the one hand, the longing has taken to imagining a perfect place: a city, for example, with a perfectly ordered spatial form. Such thinking can be seen in many places, but most recognizably in the urban plans of Howard and Le Corbusier. For Harvey, this is the triumph of spatial form over social process – a perfect society is stabilized by a perfect urban order. No (more) change required. Such ideas, Harvey argues, have become inextricably bound up with authoritarianism and totalitarianism.

And so playing with the map of spatial form has become a deathly exercise for anyone interested in democracy and freedom – and in the emancipatory city. Instead, liberation has come to be seen as a social process. On the other hand, then, Harvey identifies a utopianism of social process associated with thinkers such as Marx and Hegel. In this formula for utopia, social process determines spatial form. Social processes, as they currently stand, will have to be revolutionized, but it would be impossible to determine in advance what kinds of spatial forms might be appropriate or necessary to new forms of social relation. In this sense, spatial form is put at the service of the revolution. The down-side of this thinking is that the revolution, and its outcomes, are 'no place'. For Harvey, since these transformations remain endlessly open, they debilitate interventions in the production of space as it actually exists. For anyone wishing to bring back to life the possibility of utopian longing, there is a stark dilemma:

> Utopias of spatial form get perverted from their noble objectives by having to compromise with the social processes they are meant to control. We now see also that materialized utopias of social process have to negotiate with spatiality and the geography of place and in so doing they also lose their ideal character, producing results that are in many instances exactly the opposite of those intended. (Harvey, 2000: 179–80)

So, here's the problem with producing an emancipatory Baltimore (or any other city). On the one hand, it is comparatively easy to dream up an ideal scheme for the urban form of Baltimore – try this at home: get out a map of your favourite city and start redrawing it to make it better, freer, more democratic, just. But (and this is a big but), because it involves a democratic recognition of the city's heterogeneity, when it comes to putting the (utopian) scheme into bricks and concrete, all kinds of compromises have to be made, as the plans begin to mesh with diverse (and often contradictory) financial, political, and social interests. On the other hand, one might think that it is actually Baltimore's economic, social, and political processes that have to be transformed. Workers should be put in charge of the means of production (not that there are many 'means' of production left); new forms of political representation devised; and new kinds of social groupings developed and encouraged; even, one might add, a new settlement with Nature. But, then, the revolutionaries; are left with the problem of what to do about the physical structures of cities. Suburbs, roads, and skyscrapers are not easily torn up or torn down; nor are they easily replaced. Nor is it clear what they should be replaced with. Would utopia look like a suburb? Could it exist in a skyscraper? Such questions never find an answer and, so, nothing gets done.

Harvey's solution to this double-sided problem is to suggest that a new kind of utopianism be explored: *spatiotemporal utopianism*. This utopia engages the utopianisms of spatial form and temporal process dialectically, to create the possibility of imagining the emancipatory city as both transformative of space-time and transformed in space-time. Visionary thought would be brought to bear on both social processes and spatial form at one and the same time. The intention, then, would be to empower people to create an emancipatory city built of transformative physical forms and socialised forms of governance and economy. Each 'stage' in the freeing of the city – in its form and process – would necessitate further transformations in form and process. To get the ball rolling, Harvey suggests that empowerment might come through interventions in three areas: in government; in the economy; and in the creation of universal rights. Later in the analysis, Harvey identifies eight fronts on which political interventions can be made: transforming everything from the self to community, institutions to the built environment, ecologies to the universal principles of human action.

Inevitably, Harvey argues, such interventions carry with them the danger of authoritarianism and exclusion, but, without any interventions, the greater danger is that the catastrophe of city life will continue like this. Now is the time to begin to imagine the emancipatory city:

> There is a time and place in the ceaseless human endeavour to change the world, when alternative visions, no matter how fantastic, provide the grist for shaping powerful political forces for change. I believe we are precisely at such a moment. Utopian *dreams* in any case never entirely fade away. They are omnipresent as the hidden signifiers of our *desires*. (Harvey, 2000: 195, emphasis added)

Harvey is serious about dreams of utopia – for he uses the conceit of a dream to outline the fantastic possibility of an alternative vision. As with other utopian dreamers,[9] his vision begins in the nightmare of the present (as we have seen). And, in an appendix to *Spaces of Hope*, the dream begins to take form. In fact, the dream is not very dream-like.[10] The times and spaces of the dream are highly ordered and rational. Harvey's utopian dream begins in 2005, when global warming begins to become apparent (some might argue, this is five years after the actual event!). With growing inequalities of wealth and power, the global system comes under increasing pressure. Somewhere between 2010 and 2013, a stock market crash, originating in Russia, stretches the world financial system to breaking point. The global capitalist system goes into melt-down, provoking a military takeover in 2014. The military are violent and oppressive and, to maintain their authority, install a highly advanced system of mass surveillance. Similarly, they create global standards of governance, communication, and the like. Nevertheless, by 2019, disparate and diverse oppositional movements somehow come together to create a global revolution. Perhaps the flow of these events are kin to recent demonstrations in London. For example, on 1 May 2001, a wide variety of different anti-capitalist groups converged on Oxford Circus in London,

FIGURE 13.4   Riot police around Oxford Circus, London, 1 May 2001 (Photo: Steve Pile)

only to find themselves completely outnumbered and surrounded by police in riot gear (see Figure 13.4; see also Pile, 2003).

In David Harvey's story, the revolution is violently and ruthlessly suppressed – no surprise to those who witness the actions of militarised police forces around the world or the military policing of the world. For Harvey, though, at this point, working-class women create a global anti-military cultural movement of unstoppable force. This organisation begins in Buenos Aires and is known as 'The Mothers of Those Not Yet Born':[11] it is non-violent, and combines passive resistance with mass action. It is, ultimately, successful. And (in part two of Harvey's dream) a utopian form of global spatial organisation is installed.

It is perhaps best to think of the utopian spatial form of Harvey's dream as a kind of central place theory – this is somewhat in keeping, in fact, with the spirit of Christaller's own thinking, which had its utopian elements! The nested hierarchies of spaces and places has, at its lowest level, 'hearths' (with 20–30 adults and kids) and/or 'pradashas' (parenting collectives); then, 'neighbourhoods' (containing 10 hearths); 'edilias' (containing 200 plus neighbourhoods); 'regiona' (containing 20 to 50 edilia); 'nationa' (as a federation of regiona). It isn't that simple though, since there are no fixed spatial scales nor fixed political organizations; especially since the dream system allows for the free flow of people and goods across regiona. Here, the 'dialectic of space and time' is most apparent. The consequence of actually having no fixed scales is that the free flow of people and goods and a situation of continual political upheaval is far too dream-like – far too chaotic and disorderly. In the detail, scales are fixed, flows are controlled, exchanges between people regulated by a transparency enabled by those ever so useful military surveillance systems, and the system of political representation is completely stable (since it is, after all, democratic). In the end, the model appears to be 'utopic' because everything is collectivized: from sex and babies to goods and services – even the psyche is collectivized: individuals get psychotherapy should their passions and desires get the better of them. Possession and obsession are, in the nicest possible way, *forbidden*.

Collectives, of course, can be 'religious'; they can honour their dead, even creating festivals to pay their respects to the dead; some collectives generate new forms of 'spirit talk'. In this spirit, Harvey (2000: 278) infers that 'many of us now believe that the spirits of the dead continue to circulate among us, always'.

Now, before I begin to draw conclusions from all this, it seems worth inserting a few caveats. To be sure, Harvey warns that this dream is outlandish, outrageous and even nightmarish. His 'waking self' seems to disown it. Of course, his waking self is more interested in imagining a world without money, with free workers, without an ever-speeding up rat race pace of life, with respect and without competition (that is, with collective collaboration). These are utopianisms of process, but the dream's lesson is that these ideals must happen *somewhere*. Or else, we end up with Baltimore as it is, and the catastrophe continues. The lesson for the emancipatory city is clear: something must be done.

But Harvey's attempt to bring to life a new future remains haunted – haunted by revolutions of the past, by the violence and injustice of the present, and, ultimately, by the death of Utopia (even Hope is on the way out); the ghosts not only walk through his dream, but also through his spatiotemporal utopianism. We cannot be certain if spatial forms oppress or liberate (that is, more accurately, whom they oppress and whom they liberate, where and when). We cannot be certain that the social revolution of the twenty-first century won't be deadened by its reminiscences of, and nostalgias about, earlier revolutions. We are still not sure whether we should be letting the dead bury the dead or encouraging more 'spirit talk'.

Cities, the dilemmas of emancipation and the ghosts among us, always

In his commentary on Derrida's *Spectres of Marx*, Fredric Jameson says this:

> To forget the dead altogether is impious in ways that prepare their own retribution, but to remember the dead is neurotic and obsessive and merely

feeds a sterile repetition. There is no 'proper' way of relating to the dead and the past. (Jameson, 1999: 58–9)

It seems we are close to drawing the same conclusion, this time in relation to the emancipatory city. Thus: to emancipate the city, we must take into account the dead, but not become possessed by them. Perhaps this 'improper' haunting would appease Marx, Derrida, and Harvey. We cannot ignore the dead, otherwise we may never learn from them, nor will we honour them. But nor can we endlessly and melancholically return to the dead: lest we become unhealthily attached to them, lest we become entrapped in the relentless, drowning flow of history. If we are to emancipate the city, then we are faced with a series of dilemmas. Let me pick out two. We must liberate ourselves from the dead, but we must do so without simply denying the life of the dead: both their dreams and their nightmares. We must allow the dead to liberate us, but we must do so in ways that do not tie us to their long-gone dreams and their nightmares. The dead, as much as the living, must become part of a revolutionised history that is full of the flow of life.

Listen to Virginia Woolf (but remember Walter Benjamin too):

Where then can one go in London to find peace and the assurance that the dead sleep and are at rest? London, after all, is a city of tombs. But London nevertheless is a city in the full tide and race of human life. (Woolf, 1932: 125–6).

What is true for London, is also true for New Orleans (Roach, 1996) and Berlin (Benjamin, 1979 [1932]; Ladd, 1997; Till, 1999) and, probably, every other city. Cities, in fact, are full of the dead. They are cities of tombs. The physical structures themselves – or the gaps they leave behind – can become ghosts (see DeLyser, 1999; Kuftinec, 1998). With all these points of departure, it is easy to see why the living might require the assurance that the dead rest easy. Perhaps this is why monuments to the dead haunt the streets. On the other hand, monuments also enable the dead to cling to the city, making sure that their time is not forgotten, not passed (as Lefebvre, 1991, noted). On the other hand, exactly how monuments break into the memories of the living can be unpredictable and contested (see, for example, Johnson, 1995).

There remains an awkward accommodation between the living and the dead (and, to appease Derrida, the not-yet-born). This can be seen, just as clearly, in Harvey's own fantasy future – a fantasy which, incidentally, seems to be yet another expression of modern melancholia (see Wheeler, 1999): that is, another expression of the moderns' obsession with, and neuroses about, the need to repair the losses suffered in the wake of modernity's creative destruction and destructive creations. Ultimately, the just and free city will need to find a way to accommodate its ghosts (whether from the past, the present or the future; whether personal or collective; whether dead or alive). The lesson is already there, but it has not yet been learnt. We can see this, today, in London – the story about the Cenotaph is not yet over.

As I have said, the Cenotaph in Whitehall is the focus of an annual day of remembrance, for those who have made the ultimate sacrifice in the name of God, King or Queen, and Country. The 'day of the dead' in the year 2000 (Sunday, 12 November) was special for a number of reasons. The reinvention of Britain as an inclusive society meant that more dead people were remembered this year than previously – their ghosts, still troubled, took their place alongside the already honoured dead. Here, then, were the ghosts of the civilian dead; represented in their living form by 2,000 living civilians (the '2,000 for 2000'). Other living 'ghosts' walked: the Women's Land Army, evacuees, the ex-Services Mental Welfare Society (even ghosts need psychotherapy) and the Army Pigeon Service. (I could make much of these dead pigeons, but it is worth observing in passing that ghosts are not exclusively human affairs.) For the first time, 'Shot at Dawn' had representatives, to commemorate those who were shot – by unfriendly friendly fire – for cowardice and desertion. As Britain reinvents itself as a multicultural society, so too the dead were recognized as multicultural: representatives of the Buddhist, Hindu, Sikh, Muslim, and Greek Orthodox faiths joined Christian and Jewish leaders in laying tributes to the dead. The usually white cenotaph suddenly began to look like a rainbow: the dead, now, in full colour.

Recently, there has been bad news for the ghosts of war. The British Legion revealed that a quarter of all children had no idea what Remembrance Day stood for. These children, it seems, knew nothing of these ghosts. Days later, another threat began to haunt the Cenotaph. With cross-party support, the Labour Government proposed to build a new war memorial to members of the armed services that had died since Second World War. The plan for the memorial suggested that it would be as dramatic as America's Vietnam memorial – and, similarly, that it would name the names of the dead. The ghosts trembled and marched again. Fears were expressed that the 4,000 thus remembered would detract from the Cenotaph as the 'proper' focus of national mourning. It is not, of course, simply a question of numbers: the Cenotaph has accumulated, and continues to accumulate, more and more dead. The real danger is that the Cenotaph would become increasingly meaningless, as the dead buried their own. In this sense, the children would feel more about the 'new dead' memorial and would never appreciate the sacrifices of those previous generations embodied in the Cenotaph. The protesters of 1 May 2000, then, would be just the beginning of an exorcism of old unwanted ghosts. The second area of debate over the new memorial is as interesting: whose names would go on the monument? In other words, what would count as a 'proper' death? Surely, the Cenotaph was a flexible memorial, without names, it could remember all the dead ... and add more dead as more count as having died in the right way (including perhaps even pigeons).

Naming names would clearly be exclusive and inflexible: and this is no way to treat ghosts.

So, the lesson is there to be learnt. Cities, to be free and just, emancipatory, must be inclusive and flexible in their treatment of their ghosts. They must be prepared to add, and take away, the dead. They must be able to see themselves in relation to old dreams and old hopes, but they must not let this after-life stop them creating new dreams and new hopes. Cities must be carefully blasted from the continuum of history in order to bring about alternative histories of the future.

In fact, this may mean a decisive, flexible, and inclusive attitude to the physicality – including its spatial form – of cities. Cities can seem such spirit-less places: modern, new, secular, real. In fact, they are far from (only) so, something rational planning (whether revolutionary or not) takes no account of. In this other-worldly spirit, perhaps, there are opportunities to commemorate and accommodate the dead anew. The cenotaph, perhaps, has temporarily outlived its usefulness. Though, maybe, it could be redesignated as memorial to others who have died in armed conflict: perhaps commemorating those killed by the British in its Imperial adventures; or, perhaps it might celebrate the victories of anti-colonial struggles.[12] Alternatively, we might think of making sacred some rather profane places – like railway stations: Euston could become a memorial to those who have died making, and running, railways; the Cross in King's Cross could be a memorial to all those passengers who have died (or felt crucified by lost time). It might not end there: the Millennium Dome could be a monument to the dead ideals of New Labour's New Britain (hmm: perhaps it already is). If this seems fanciful, then we should remember that this reinvention of the places of the dead (and the living) is already underway.

In 2000, the fourth plinth in Trafalgar Square did not have a dead (white male) war hero on top of it (as have the other three), but a statue of a tree with its roots wrapped around a book that is pressing down on a human face. The title of Bill Woodrow's piece was 'Regardless of History'. For Woodrow, the sculpture signalled the supremacy of Nature. Alternatively, it could also say something about the roots of knowledge and their entanglement in the past, while the tree stretches for the sky, towards the future – regardless of its history, but also rooted in it. It is time for the city, itself, to learn from its ambivalence towards the dead. And to create its own poetry from the future *and* the past.

Notes

1 Protests also occurred outside the capital. Manchester also had its May Day protests, with some 200 *Reclaim the Streets* activists attempting to storm the Arndale Centre.

2 Some speculated that as many as 30,000 officers were present or on call in central London that day.

3 Cited in the *Guardian*, 2 May 2000, front page.

4 It is not mere rhetoric to suggest that, months after September 11, the ghosts of New York haunt the world. The *Guardian Weekend*, 29 December 2001, for example, described the ruins of The World Trade Center as 'ghostly' (page 30) – an idea that is firmly fixed in descriptions of post-September 11 New York and New Yorkers.

5 TIKB is the Turkish Communist Workers Association. The TIKB had been one of the most prominent groups at the peaceful May Day demonstration in 1999. This demonstration was notable for its anti-racist and anti-homophobic message in the wake of David Copeland's three nail-bomb attacks in April that year (see Pile, 2003).

6 This social memory of Churchill was invoked again in the debate over Britain's participation in the European Rapid Response Force. His statue, with Big Ben in the background, was used by television news programmes (for example *Newsnight*, 22 November 2000) to dramatize the long shadow that the ghosts of the Second World War cast over contemporary Britain and its attitudes towards Europe.

7 Much has been made of ghosts being in a time 'out of joint'. This largely stems from Derrida's (1993: xix, 17) discussion of Hamlet. See also Jameson (1999) and Laclau (1995).

8 For a related diagnosis, see Mooney et al. (1999).

9 Here, I am thinking particularly of Walter Benjamin (Pile, 2000). Nevertheless the dream is a common device for enabling other possible worlds to be imagined.

10 On dreams, see Pile (1998; forthcoming).

11 There are other ghosts here, see Radcliffe (1993).

12 Remember, then, that the IRA launched mortar bombs from just behind the Cenotaph. Their target was John Major and his cabinet, who were meeting at Number 10 Downing Street. No-one was killed, but British pride was severely wounded.

References

Bellamy, E. (2001) *Looking Backward, 2000–1887*. Harmondsworth: Penguin.

Benjamin, W. (1979 [1935]) 'A Berlin chronicle', in W. Benjamin (1979) *One Way Street and Other Writings*. London: Verso. pp. 293–346.

Benjamin, W. (1999 [1935]) 'Paris, the capital of the nineteenth century', in W. Benjamin *The Arcades Project*. Cambridge, MA: Harvard University Press. pp. 3–13.

DeLyser, D. (1999) 'Authenticity on the ground: engaging the past in a California ghost town', *Annals of the Association of American Geographers*, 89: 602–32.

Derrida, J. (1993) *Spectres of Marx: The State of the Debt, the Work of Mourning, and the New International*. London: Routledge.

Guardian (2000) 'Protests Erupt in Violence', 2 May, front page.

Harvey, D. (1973) *Social Justice and the City*. London: Edward Arnold.

Harvey, D. (2000) *Spaces of Hope*. Edinburgh: Edinburgh University Press.

Howard, E. (1902) [1965] *Garden Cities of Tomorrow*. London: Faber and Faber.

Jameson, F. (1999) 'Marx's purloined letter', in M. Sprinker (ed.), *Ghostly Demarcations: A Symposium on Jacques Derrida's Spectres of Marx*. London: Verso. pp. 26–67.

Johnson, N. (1995) 'Cast in stone: monuments, geography and nationalism', *Environment and Planning D: Society and Space*, 13: 51–66.

Kuftinec, S. (1998) '[Walking through a] Ghost town: cultural hauntologie in Mostar, Bosnia-Herzegovina or Mostar: a performance review', *Text and Performance Quarterly*, 18: 81–95.

Laclau, E. (1995) 'The time is out of joint', *Diacritics*, 25: 86–96.

Ladd, B. (1997) *The Ghosts of Berlin: Confronting German History in the Berlin Landscape*. Chicago, IL: University of Chicago Press.

Lefebvre, H. (1991) *The Production of Space*, trans. D. Nicholson-Smith. Oxford: Basil Blackwell.

Marx, K. (1852 [1973]) 'The Eighteenth Brumaire of Louis Bonaparte', in K. Marx, *Surveys from Exile*. Harmondsworth: Penguin. pp. 143–249.

Mooney, G., Pile, S. and Brook, C. (1999) 'On orderings and the city', in S. Pile, C. Brook and G. Mooney (eds), *Unruly Cities? Order/Disorder*. London: Routledge in association with the Open University. pp. 345–63.

Pile, S. (1998) 'Freud, dreams and imaginative geographies', in A. Elliott (ed.), *Freud 2000*. Cambridge: Polity Press. pp. 204–34.

Pile, S. (2000) 'Sleepwalking in the modern city: Walter Benjamin and Sigmund Freud in the world of dreams', in S. Watson and G. Bridge (eds), *Blackwell Companion to Urban Studies*. Oxford: Basil Blackwell. pp. 75–86.

Pile, S. (2003) 'Struggles over geography', in K. Anderson, M. Domosh, S. Pile, and N. Thrift (eds), *Handbook of Cultural Geography*. London: Sage. pp. 23–30

Pile, S. (forthcoming) 'Worlding dreams: space, psychoanalysis and the city', in N. Leach (ed.), *Psychoanalysis and Space*.

Radcliffe, S. (1993) 'Women's place/el lugar de mujeres: Latin American and the politics of gender identity', in M. Keith and S. Pile (eds), *Place and the Politics of Identity*. London: Routledge. pp. 102–16.

Roach, J. (1996) *Cities of the Dead: Circum-Atlantic Performance*. New York: Columbia University Press.

St. Clair, J. (1999) 'Seattle diary: it's a gas, gas, gas', *New Left Review*, 238: 81–96.

Thomson, J. (1880) [1998] *The City of Dreadful Night*. Edinburgh: Canongate Books.

Till, K. (1999) 'Staging the past: landscape designs, cultural identity and *Erinnerungspolitik* at Berlin's *Neue Wache*', *Ecumene*, 6: 251–83.

Wheeler, W. (1999) *A New Modernity? Change in Science, Literature and Politics*. London: Lawrence & Wishart.

Williams, R. (1989) *Resources of Hope*. London: Verso.

Woolf, V. (1932) 'The London scene: abbeys and cathedrals', in V. Woolf (1993) *The Crowded Dance of Modern Life. Selected Essays: Volume 2*. Harmondsworth: Penguin. pp. 122–7.

Young, H. (2000) 'There is a gap in the market for serious radicalism', *Guardian* (2000), 2 May, p. 18.

Reflections

14 The 'Emancipatory' City?

Ash Amin and Nigel Thrift

In the long lineage of Western thought on urban life the city has very often been seen as a forcing ground for a politics of emancipation. Thus, the classical Graeco-Roman city is where the rule of democracy is supposed to have arisen, a democracy based upon the public deliberations of a supposedly 'free' citizenship, supported by a variety of uniquely urban political institutions. The medieval city, and later, the Renaissance city are held responsible for such seminal events as the rise of guild politics, the forging of institutions of civic republicanism and the principle of sanctuary based around the rise of independent city states. The Enlightenment city – through its institutions of learning, intellectual exchange, and secular science – is associated with the rise of universalism and a cosmopolitan ethos. And so on.

Throughout the course of modern history, in other words, the city has been seen as in the vanguard of politics, drawing upon resources like the murmur of the crowd, mass discontent, the constant flow of outsiders, and the sheer weight of juxtaposed difference and supported by a powerful infrastructure of intellectuals, activists, and dissidents, spaces of gathering and organization, diverse forms of association, and varied technologies of representation and communication. It is no surprise, therefore, that the city has featured centrally in imaginaries of modern utopia and designs for the good life, whether these are constructed around garden cities, ideal civic spaces, futurist techno-conquests of nature, managed movement and engagement, or homes fitted out as havens for heroes of consumption.

The common thread running through this lineage is the interpretation of the political through the city: the urban is the formative arena of both the 'major' politics of grand schemes and protest movements and the 'minor' politics of everyday struggle and discontent. But how far the contemporary city even remotely resembles this understanding of a politics ventured by and through the urban is a question well worth asking.

To begin with, the grand designs of emancipation based upon the mobilization of equally grand urban processes of some form or another have been buffeted by a creeping barrage of dystopian imaginaries of the city playing on the anomie, egotism, superficiality, criminality, poverty, terror, and violence said to characterize modern urban life. If any putative association with utopia still lingers, it does so in the shape of a 'false' or ephemeral politics of capitalist contentment based upon the seductions of urban life – the spectacle of city glitter, the specious fulfilments of urban consumption, and the self-absorptions of the urban lifestyle. Then, emancipatory politics with a big 'P' are no longer a particularly urban affair. The institutions of politics have gone national and global while social movements snake their way through an array of networked spaces. Thus the public arena and public culture are no longer reducible to the urban (owing to the rise of virtual and distanciated networks of belonging and communication). The political can no longer (if it ever could) be collapsed into a simple notion of the territorial, and certainly not the urban territorial. There does not seem to be a special place for the powers of urban citizenship, urban institutions, urban space, urban mingling, urban association.

What, then, remains of the urban in the politics of emancipation? Has the political shifted to another scale altogether or to a relational register made up of varied, partially connected spatial expressions? It would be very easy to assert that not very much remains of the urban as a political force and to side either with the dystopians or with those urban visionaries nostalgic for the original political effects of urban aggregation (from polis, public deliberation and agora to sociality, sanctuary and sedition). But we want to make a different argument.

We want to recognize in the urban a stuttering but nevertheless vocal potentiality thrown up, first, by the involuted character of the city arising out of its role as a confluence of flows and difference (Massey, 1999) and, second, by the constant hum of the everyday and prosaic web of practices that makes the city into such a routinely frenetic place (Amin and Thrift, 2002). The result, we would argue, is the existence of a politics of the minor register – forever changing, always fragmentary, full of small gains and losses that never quite add up, embroiled in many spatial circuits, and 'political' but in non-traditional ways. Such a politics is not of minor significance, however. Its struggles dance inventively with small things – gestures, encounters, noise, smells, relationships, codes, rules and conventions, walks, car journeys, friendships, fears, daily joys and irritations – that can form larger political proclivities and, as significantly, the political itself as a field of formation, action and engagement.[1] But there is more. The modern city is so continuously in movement and, consequently, so full of unexpected interactions that all kinds of spatialities are continually being opened up that provide the resources for continuing political invention.

All we have space for in this short closing intervention are some illustrations of our argument. First, the city continues to provide a site where different flows and things come together in alignments from which surprising juxtapositions can be produced, so that it is possible to think of a politics of propinquity. To be clear, such a politics is not reducible to local influences and local outcomes, not least because practices and prospects are constituted in and through relational networks of varying spatial reach that constantly outrun descriptions like large or small (see Latour, 2002). Rather, it is *instanciated* by spatial juxtaposition. For example, the concentration in global cities of transnational elites in close proximity to transnational migrants and dispossessed peoples has spawned awareness of both a new politics of inequality as well as new strategies of domination and resistance made up of distributed fragments, global alliances, and local resources (Sassen, 1999). Urban multiculturalism is forging new senses of place that draw on a lightly-drawn mix of local commonalities and shared belongings but which still have enough depth to allow citizens in one city or parts of a city to enter into strategic political alliances. Spatial context can make all the difference, as is so evident from the microgeography of the politics of race and inter-ethnic relations, which shows that even adjacent neighbourhoods in a city can yield drastically different attitudes and practices towards minorities and strangers (Amin, 2002).

Second, the everyday city is overflowing with performative political spaces that interact and interdigitate with each other. For instance, there are the politics involved with the frames surrounding everyday life, simple elements such as standards, metrics, software programmes, and other cursive technologies, which are all the more powerful for being so rarely questioned, but which daily reinvent the political (Thrift, 2004; Thrift and French, 2002). We might also point to a politics of the micro-spaces of the city. Over the years feminists, in particular, have foregrounded spaces of the domestic and the often quite vicious power plays acted out within them (Watson, 2002). Similarly, there is a whole politics of embodiment, from the minutiae of gesture to the movement patterns of the crowd, that still has only rarely been explored systematically (Finnegan, 2002; Sennett, 1970). Finally, the politics of turf remains ever strong, around disputes over who has rights to public spaces, conflicting views on the dividing line between the domestic and the public sphere, planning conflicts arising from varying public demands, and continuing warring between claimants of a congested space.

Third, such a politics brings into play means of sensing the city that have rarely been recognized as weapons of emancipation and political struggle. And yet, modern cities, because of their vitality and absorptive capacity, are full of representational experiments with and new inventions of these sensings that count as a kind of politics: all the way from the many children of Surrealism, through to the recent work by numerous teenagers on the social possibilities of wireless communication. In turn, this extraordinary resource shows the importance of ethico-aesthetic invention as a key moment in a revitalized urban politics (Guattari, 1996). For example, collage, graffiti, and other modes of aethesticization, have now become a standard means of urban communication (Lash, 2001), as have the 'marvellous worlds' triggered by the many objects of enchantment (from psychotropic drugs to moments of joy in thrall to an urban 'sublime') that sustain hybrid cultural crossings, a politics of possibility, and new framings of embodiment, interpretation, and engagement (Bennett, 2001).

To conclude, much of what is specific and important about the modern city's political force is not concerned with participation in politics with a conventional capital P. This is not, of course, an argument against participation in such politics. Rather, it is an argument for considering new forms of molecular politics that not only vie for public attention with politics as conventionally understood but whose concerns very often act as a wellspring for driving wider social and cultural change. It is also about taking more seriously the minor forms of association that abound in the city and allowing these kinds of politics to subsist (Joseph, 2002). In discussions of emancipatory politics, we need to reconsider the numerous forms of ordinary urban sociality. Somehow they have come to be looked down upon as not worthy of the ascription 'political', even though most political acts precisely draw upon that everyday sociality. Take one of the constant themes of Graeco-Roman thought, friendship, as an instance. Friendship and friendship-based associations have become an increasingly important element of the urban social glue, many of whose pleasures lie in simply relating to others. Even though the forging of the bonds of friendship may be the result of the increasing emphasis on relationship as a value in itself, such bonds also take us back to the very roots of cities as sites of association, and through this, political organization. Thus what may seem routine, even trivial, may have all manner of political resonances that we are only just beginning to understand – and mobilize.

Note

1 However, it is important to note here that we are not subscribing to the 'evasive everydayness' model so beloved of large parts of cultural studies which seems to us deeply flawed (see Morris, 1998).

References

Amin, A. (2002) 'Ethnicity and the multicultural city: living with diversity', *Environment and Planning A*, 34: 959–80.

Amin, A. and Thrift, N. J. (2002) *Cities: Reimagining the Urban*. Cambridge: Polity Press.

Bennett, J. (2001) *The Enchantment of Modern Life*. Princeton, NJ: Princeton University Press.

Finnegan, R. (2002) *Communicating: The Multiple Modes of Human Interconnection*. London: Routledge.

Guattari, F. (1996) *Chaosmosis: An Ethico-Aesthetic Paradigm*. Sydney: Power Publications.

Joseph, M. (2002) *Against the Romance of Community*. Minneapolis: University of Minnesota Press.

Lash, S. (2001) *Critique of Information*. London: Sage.

Latour, B. (2002) 'Gabriel Tarde and the end of the social', in P. Joyce (ed.), *The Social in Question: New Bearings in History and the Social Sciences*. London: Routledge. pp. 117–32.

Massey, D. (1999) 'Cities in the world', in D. Massey, J. Allen, J. and S. Pile (eds), *City Worlds*. London: Routledge. pp. 99–156.

Morris, M. (1998) *Too Soon Too Late: History in Popular Culture*. Bloomington, IN: Indiana University Press.

Sassen, S. (1999) 'Globalization and the formation of claims', in J. Copjec and M. Sorkin (eds), *Giving Ground: The Politics of Propinquity*. London: Verso. pp. 86–105.

Sennett, R. (1970) *Flesh and Stone*. New York: Norton.

Thrift, N. J. (2004) 'Remembering the technological unconscious by foregrounding knowledges of position', *Environment and Planning D: Society and Space*, 22: 175–90.

Thrift, N. J. and French, S. (2002) 'The automatic production of space', *Transactions of the Institute of British Geographers*, 27: 309–35.

Watson, S. (2002) 'The public city', in J. Eade and C. Mele (eds), *Understanding the City*. Oxford: Blackwell. pp. 49–65.

15 The Right to the City

David Harvey

The city, the noted urban sociologist Robert Park once wrote, is:

> Man's most consistent and on the whole, his most successful attempt to remake the world he lives in more after his heart's desire. But, if the city is the world which man created, it is the world in which he is henceforth condemned to live. Thus, indirectly, and without any clear sense of the nature of his task, in making the city man has remade himself. (1967: 3)

The right to the city is not merely a right of access to what already exists, but a right to change it after our heart's desire. We need to be sure we can live with our own creations (a problem for every planner, architect and utopian thinker). But the right to remake ourselves by creating a qualitatively different kind of urban sociality is one of the most precious of all human rights. But the sheer pace and chaotic forms of urbanization throughout the world have made it hard to reflect on the nature of this task. We have been made and re-made without knowing exactly why, how, wherefore and to what end. How then, can we better exercise this right to the city?

The city has never been a harmonious place, free of confusions, conflicts, violence. Only read the history of the Paris Commune of 1871, see Scorsese's fictional depiction of *The Gangs of New York* in the 1850s, and think how far we have come. But then think of the violence that has divided Belfast, destroyed Beirut and Sarajevo, rocked Bombay, even touched the 'City of Angels'. Calmness and civility in urban history are the exception not the rule. The only interesting question is whether the outcomes are creative or destructive. Usually they are both: the city is the historical site of creative destruction. Yet the city has also proven a remarkably resilient, enduring and innovative social form.

But whose rights and whose city? The communards of 1871 thought they were right to take back 'their' Paris from the bourgeoisie and imperial lackeys. The monarchists who killed them thought they were right to take back the city in the name of God and private property. Both catholics and the protestants thought they were right in Belfast as did Shiv Sena in Bombay when it violently attacked muslims. Were they not all equally exercising their right to the city? 'Between equal rights', Marx (1973: 344) once famously wrote, 'force decides'. So is this what the right to the city is all about? The right to fight for one's heart's desire and liquidate anyone who gets in the way? It seems a far cry from the universality of the UN Declaration on Human Rights. Or is it?

Marx, like Park, held that we change ourselves by changing our world and vice versa. This dialectical relation lies at the root of all human labour. Imagination and desire play their part. What separates the worst of architects from the best of bees, he argued, is that the architect erects a structure in the imagination before materializing it upon the ground. We are, all of us, architects, of a sort. We individually and collectively make the city through our daily actions and our political, intellectual and economic engagements. But, in return, the city makes us. Can I live in Los Angeles without becoming a frustrated motorist?

We can dream and wonder about alternative urban worlds. With enough perseverance and power we can even hope to build them. But utopias these days get a bad rap because when realized they are often hard to live with. What goes wrong? Do we lack the correct moral and ethical compass to guide our thinking? Could we not construct a socially just city?

But what is social justice? Thrasymachus in Plato's *Republic* (1941: 18) argues that 'each form of government enacts the laws with a view to its own advantage' so that 'the just is the same everywhere, the advantage of the stronger'. Plato rejected this in favour of justice as an ideal. A plethora of ideal formulations now exist. We could be egalitarian; utilitarian in the manner of Bentham (the greatest good of the greatest number); contractual in the manner of Rousseau (with his ideals of inalienable rights) or John Rawls; cosmopolitan in the manner of Kant (a wrong to one is a wrong to all); or just plain Hobbesian, insisting that the state (Leviathan) impose justice upon reckless private interests to prevent social life being nasty, brutish and short. Some even argue for local ideals of justice, sensitive to cultural differences. We stare frustratedly in the mirror asking: 'which is the most just theory of justice of all?' In practice, we suspect Thrasymachus was right: justice is simply whatever the ruling class wants it to be.

Yet we cannot do without utopian plans and ideals of justice. They are indispensable for motivation and for action. Alternative ideas coupled with outrage at injustice have long animated the quest for social change. We cannot cynically dismiss either. But we can and must contextualize them. All ideals about rights hide suppositions about social processes. Conversely, social processes incorporate certain conceptions of rights. To challenge those rights is to challenge the social process and vice versa. Let me illustrate.

We live in a society in which the inalienable rights to private property and the profit rate trump any other conception of inalienable rights you can think of. This is so because our society is dominated by the accumulation of capital through market exchange. That social process depends upon a juridical construction of individual rights. Defenders argue that this encourages 'bourgeois virtues' of individual responsibility, independence from state interference, equality of opportunity in the market and before the law, rewards for initiative, and an open market place that allows for freedoms of choice. These rights encompass private property in one's own body (to freely sell labour power, to be treated with dignity and respect and to be free from bodily coercion), coupled with freedoms of thought, of expression and of speech. Let us admit it: these derivative rights are appealing. Many of us rely heavily upon them. But we do so much as beggars live off the crumbs from the rich man's table. Let me explain.

To live under capitalism is to accept or submit to that bundle of rights necessary for endless capital accumulation. 'We seek', says President Bush (2002) as he goes to war, 'a just peace where repression, resentment and poverty are replaced with the hope of democracy, development, free markets and free trade'. These last two have, he asserts, 'proved their ability to lift whole societies out of poverty'. The United States will deliver this gift of freedom (of the market) to the world whether it likes it or not. But the inalienable rights of private property and the profit rate (earlier also embedded, at US insistence, in the UN declaration) can have negative even deadly consequences.

Free markets are not necessarily fair. 'There is', the old saying goes, 'nothing more unequal than the equal treatment of unequals'. This is what the market does. The rich grow richer and the poor get poorer through the egalitarianism of exchange. No wonder those of wealth and power support such rights. Class divisions widen. Cities become more ghettoized as the rich seal themselves off for protection while the poor become ghettoized by default. And if racial, religious and ethnic divisions crosscut, as they so often do, with struggles to acquire class and income position, then we quickly find cities divided in the bitter ways we know only too well. Market freedoms inevitably produce monopoly power (as in the media or among developers). Thirty years of neoliberalism teaches us that the freer the market the greater the inequalities and the greater the monopoly power.

Worse still, markets require scarcity to function. If scarcity does not exist then it must be socially created. This is what private property and the profit rate do. The result is much unnecessary deprivation (unemployment, housing shortages, etc.) in the midst of plenty. Hence the homeless on our streets and the beggars in the subways. Famines occur in the midst of food surpluses.

The liberalization of financial markets has unleashed a storm of speculative powers. A few hedge funds, exercising their inalienable right to make a profit by whatever means rage around the world, speculatively destroying whole economies (such as that of Indonesia and Malaysia). They destroy our cities with their speculations, reanimate them with their donations to the opera and the ballet while, like Kenneth Lay of Enron fame, their CEOs strut the global stage and accumulate massive wealth at the expense of millions. Is it worth the crumbs of derivative rights to live with the likes of Kenneth Lay?

If this is where the inalienable rights of private property and the profit rate lead, then I want none of it. This does not produce cities that match my heart's desire, but worlds of inequality, alienation and injustice. I oppose the endless accumulation of capital and the conception of rights embedded therein. A different right to the city must be asserted. Those that now have the rights will not surrender them willingly. 'Between equal rights, force decides'. This does not necessarily mean violence (though, sadly, it often comes down to that). But it does mean the mobilization of sufficient power through political organization or in the streets if necessary to change things. But by what strategies do we proceed?

No social order, said Saint-Simon, can change without the lineaments of the new already being latently present within the existing state of things. Revolutions are not total breaks but they do turn things upside down. Derivative rights (like the right to be treated with dignity) should become fundamental and fundamental rights (of private property and the profit rate) should become derivative. Was this not the traditional aim of democratic socialism?

There are, it turns out, contradictions within the capitalist package of rights. These can be exploited. What would have happened to global capitalism and urban life had the UN declaration's clauses on the derivative rights of labor (to a secure job, reasonable living standards and the right to organize) been rigorously enforced? But new rights can also be defined: like the right to the city which, as I began by saying, is not merely a right of access to what the property speculators and state planners define, but an active right to make the city more in accord with our heart's desire, and to re-make ourselves thereby in a different image.

But this can never be a purely individual affair. It demands a collective effort. The creation of a new urban commons, a public sphere of active democratic participation, requires that we roll back that huge wave of privatization that has been the mantra of a destructive neoliberalism. We must imagine a more inclusive, even if continuously fractious, city based not only upon a different ordering of rights but upon different political-economic practices. If our urban world has been imagined and made then it can be re-imagined and re-made. The inalienable right to the city is worth fighting for. 'City air makes one free', it used to be said. The air is a bit polluted now. But it can always be cleaned up.

References

Bush, G.W. (2002) 'Securing freedom's triumph', *New York Times*, 11 September, p. A33.

Marx, K. (1973) *Capital, Volume 1*, trans. B. Fowkes. New York: Vintage.

Park, R. (1967) *On Social Control and Collective Behavior*. Chicago, IL: Chicago University Press.

Plato (1941) *The Republic*, trans. F. M. Cornford. Oxford: Oxford University Press.

Index